# NAVIGATING UNCERTAINTY

## Simple Scientific Insights for Everyday Life

**AUTHOR**

**Anna L. Pereira**    PhD, MS

**ACADEMIC ADVISOR**

**Karl J. Friston**    MBBS, MA, MRCPsych, MAE, FMedSci, FRSB, FRS

**DESIGN & ILLUSTRATIONS**

**Ludovica Cestarelli**

**EDITOR**

**Zach Daugherty**

**REVIEWER & REFERENCES**

**Yaqi (Sylvia) Zhang**    MSc

1st Edition 2025

ISBN: 979-8-9933192-0-9

With gratitude to the researchers, scientists, practitioners, spiritual leaders, and teachers whose insights are woven into these pages.

# FOREWORD
## by Karl J. Friston

When invited to write the foreword for this book, I was delighted to accept. When asked to make it personal, I smiled: I was reminded of my usual quip when asked why I became a neuroscientist. My usual answer is "because I want to study the most interesting thing in the universe; namely, me". This somewhat egocentric answer sets the stage for a much deeper insight—pursued in the pages that follow—**that to understand the self is to understand others.** One can read this at a number of levels, ranging from how we make sense of our lived world through to the choices we make as parents, colleagues, and citizens.

I think of myself as a good citizen. But what does that mean? What does it mean to be good? I could answer that I spend my life trying to understand how the brain works and—as a clinical scientist—understand the mechanisms that underwrite the psychopathology and distress caused by brain disorders. The dénouement (outcome) of this endeavour—in my world—can be cast as a physics of sentience, a mechanics of the mind, known variously as the free energy principle, or its applications such as predictive processing and active inference.

But why appeal to physics to understand the mindful processes entailed by our sense-making and decision-making? The answer is that physics provides a precise and formal set of methods that can be used to reconstitute, reproduce,

**DEFINITION**

The study of mental health challenges.

or replicate the processes that make us think and behave in the way that we do. Interestingly, the notion of 'self' is the very first thing to emerge in any physics. For example, the physics of self-organisation, in open systems, foregrounds the notion of 'self as a system' that is open to exchange with the world. This begs a fundamental question: what is the 'self' in self-organisation? The answer is a bit technical but revealing. It means that you have to distinguish self from non-self and, in so doing, enshroud the self in something called a Markov blanket. The technical details are not important, but the physics of Markov blankets tells us something quite interesting: the very existence of this separation— i.e., possessing a Markov blanket—means that you can always be described as striving to minimise something called self-information. So what is self-information? It is simply a measure of how surprising a particular exchange with the world would be, given the kind of thing that you are. Put simply, this means that to be a 'self' is to self-organise in a way that minimises self-information (a.k.a., free energy). In other words, you avoid surprising, uncharacteristic exchanges—through your Markov blanket—with the sensed world.

But sensory exchange—what we see, hear, and feel—is not the complete story. There is a two-way traffic across your Markov blanket, in which the world impresses itself upon you—in the form of sensory input—while you impress yourself on the world through acting upon that world. In brief, this means that to exist (i.e., to be a self) is to act in a way that minimises the surprise you expect, consequent upon that action, choice, or decision. So, what is expected surprise? **It is just uncertainty.** This means that things like you and me are compelled—by

**ANNOTATION**

Minimizing self-information means reducing unexpected events that could threaten stability, not limiting self-knowledge.

our very existence—to resolve uncertainty about the world in which we find ourselves. In short, to exist is to be curious.

A revealing interpretation—of this physics of selfhood—rests upon another view of self-information or surprise. This interpretation inherits from probability theory, often cast in terms of something called Bayesian mechanics or inference. The technical details are, again, unimportant, but they furnish a compelling reinterpretation of ourselves as curious creatures. This follows from the fact that one can express surprise as the negative evidence for our internal or world models that we used to predict the consequences of our actions. These are called generative models because they generate predictions of what we encounter now or in the future, given our beliefs about current states of affairs. On this view, minimising self-information or surprise just is maximising the evidence for our generative models of the world. In turn, this leads to a particular (Bayesian) mechanics that is apt to describe sense-making and decision-making in a way that discloses the fundamentals of motivated behaviour: the curiosity above now translates into epistemic imperatives that dominate most of what we do in securing (i.e., maximising) evidence for our generative models. This picture of what it is to be a self—as gathering evidence for our models of the lived world—is summarised neatly as self-evidencing. This book is about self-evidencing, namely, gathering evidence for my model of the world, which includes you. The corollaries of this kind of self-evidencing lead to mutual understanding, as evinced by notions of communication, coherence, and compassion. But where does my model of you come from?

**ANNOTATION**

In other words, our curiosity pushes us to keep testing and updating our mental models so they line up better with reality.

**DEFINITION**

A way of staying alive by keeping your inner sense of things matched to reality.

Imagine being born into the world with no experience, other than the warm, muted interoceptive (internal) and exteroceptive (external) sensations afforded by a mother's womb. To self-evidence, you now have to build a generative model that best explains the flood of sensations on entering the world. Happily, many mothers and other caregivers make this job as easy as possible by providing the right kind of sustaining and nurturing sensations that enable you to learn about your world. The first thing you might notice is that your sensory fluctuations are best explained by a qualitative difference between when your mother is present and when she is not. This enables you to start 'carving nature at its joints' and install—in your generative model—the notion of 'mother' as an autonomous (independent) cause of your sensations; namely, mother is a 'self' in her own right. Are there other things? And, crucially, "am I a self?". Arguably, this is the genesis of selfhood (i.e., my 'self') that supervenes on the notion of an archetypal 'self'; namely, [m]other.

In this sense, we come back to the notion that to understand oneself is to understand others and, in later life, vice versa: equipped with the notion of 'self' one can now explain—as the days and weeks pass—your neonatal world in terms of many 'others' that are, perhaps, experiencing the world as you experience it. In psychology, this is referred to as 'theory of mind' and could, arguably, underwrite selfhood through the ability to disambiguate the causes of your sensations through attribution of agency. **The notion that you may experience the world in the same way that I experience the world has all sorts of wonderful benefits for self-evidencing.** For example, inferring

**ANNOTATION**

In other words, my sense of self arises from, and depends on, first recognizing a caregiver as a separate self.

**DEFINITION**

Relating to newborn children.

**ANNOTATION**

In other words, theory of mind is the ability to imagine what someone else might be thinking or feeling.

your intentions on the basis of what I see you doing by asking "what intentions would lead me to behave like you?". One can extend these arguments into the philosophical domain and start to ask questions about consciousness, and to what extent being conscious rests upon self-evidencing in a world constituted—and co-constructed—by other things like me. These and other deep questions are unpacked in the pages that follow.

In closing, I return to my personal mandate. When applying to university, Cambridge was the only place that offered the opportunity to study both physics and psychology as part of medical training. I therefore earnestly studied the entrance requirements and came across an example essay question that intrigued me then, and ever since: *"Primary ontological security: discuss"*. **I had no idea what 'primary ontological security' meant (and am still unsure). However, I think this book is the closest thing I have seen to a comprehensive and compelling answer.**

**DEFINITION**

A basic sense of safety in simply existing. The feeling that the world is stable, predictable, and that you have a secure place in it.

**Karl J. Friston**
MBBS, MA, MRCPsych, MAE, FMedSci, FRSB, FRS

**Professor:** Queen Square Institute of Neurology, University College London
**Honorary Consultant:** The National Hospital for Neurology and Neurosurgery

# TABLE OF CONTENTS

**CHAPTER 1**   **Finding Harmony Amongst Uncertainty**   1

**CHAPTER 2**   **A Story: How This Book Came Into Being**   7

**CHAPTER 3**   **Understanding Ourselves Through Physics**   13

The Role of Science in Clarity

From Uncertainty to Understanding

Active Inference and Free Energy

**CHAPTER 4**   **Charting Your Course**   19

**CHAPTER 5**   **The Active Inference Cycle: Our Lived Experience**   23

**CHAPTER 6**   **The Free Energy Principle: Seeking Harmony & Dissonance**   31

**CHAPTER 7**   **Homeostasis: A Foundation of Life**   39

**CHAPTER 8**   **Self-Knowing: Turning Up the Light**   45

Deepening Self-Knowledge Through Action

The Art of Balancing Precision

Meditation: One Gateway to Balanced Precision

**CHAPTER 9**   **Adjusting the Zoom and Focus of Our Awareness**   57

A Simplified Explanation of Attention and Precision

**CHAPTER 10**     **The Markov Blanket: A Comforting Concept**    65

**CHAPTER 11**    **Uncertainty as Opportunity: Benefits of the Unknown**    73

**CHAPTER 12**    **An Empathetic Lens: How Surprise & Uncertainty Affect Connection**    79

Discomfort of the Unexpected

Sensory Attenuation

The Narrowing Effect of Threat and Uncertainty

Tying Back in Empathy

**CHAPTER 13**    **Intentions and the Bidirectional Nature of the Brain**    93

**CHAPTER 14**    **Willpower vs. Volition: A More Effective Use of Our Finite Energy**    103

Understanding the Role of Volition

The Physiology of Misalignment

Self-Efficacy: A Tool for Volition

Balancing Pragmatic and Epistemic Intentions

Leveraging the Entire System

Moving Toward Alignment

**CHAPTER 15**    **Embodiment: A Cohesive Experience**    113

**CHAPTER 16**    **Play: Reducing Uncertainty Through Joy**    121

Tiny Scientists at Work

The Social Side of Play

Reconnecting with Play

**CHAPTER 17**    **A Playground of Possibilities: Discovering a New Place Within**    129

**CHAPTER 18**    **Rest and the Gifts of Intuition**    135

The Science Behind Rest

Sleep as Neural Housekeeping

Moment-to-Moment Rest

The Survival Perspective

Personal Reflections on Rest

Practical Implications

**CHAPTER 19**    **The Artist Within the Active Inference Cycle**    143

Two Scenes, One Walk

Curator of Action

Curator of Attention

Sticky States

The Artist's Opportunity

**CHAPTER 20**    **Laying Out Our Palette: Getting to Know Our Volition**    153

Choosing Our Brushes: Four Possible Contrasts

Expanding Our Palette

Becoming Skilled Artists

**CHAPTER 21**    **The Value of Love: A Guide for Understanding and Connection**    165

Love as an Inference Process

Romantic Love

Friendships

Self-Love

Love as an Expansion of the Self

The Role of Sensory Attenuation in Love

Cultivating Love in Our Lives

**CHAPTER 22**   **Deepening Self-Knowledge: Understanding the Mind's Layers**          181

Self-Love as a Gateway for Self-Knowing

The Layers of the Mind

When Regulation Fails: The Challenges of Misalignment

Some Practical Tools: Meditation, Dream Journaling, and Therapeutic Approaches

A Window into Psychedelic Therapies

When Others Choose a Different Path

Self-Knowing and the Shape of Loneliness

A Quick Note on Epigenetics

Closing Thoughts: Choosing Our Path Forward

**CHAPTER 23**   **Safe Containers: The Foundation for Growth and Connection**          201

What Are Safe Containers?

The Benefits of Cultivating Safe Containers with Others

Safe Containers at Different Scales

Somatic Awareness and Self-Regulation

Communication: One-Way Safe Containers Are Reinforced or Broken

Touch: The Subtle Power of Physical Presence

Actions: How We Create or Undermine Safe Containers

Safe Containers and Neurological Development

Balance in Creating Safe Containers

The Role of Epistemic Trust in Safe Containers

The Limitation of Trying to Change Others

The Power of Repair: How Safety Grows Through Imperfection

Building Internal Stability in Unstable Times

Attachment Theory Through the Lens of Active Inference

**CHAPTER 24**  **Free Energy in the System: Why Othering Hurts Us All**  225

Personal Story: Portugal

Applied to Active Inference and Free Energy

What the Science Suggests

**CHAPTER 25**  **Neurodiversity—A Valuable Human Construct**  233

Neurodiversity Through the Lens of Active Inference

Some Types of Neurodivergency

Autism: A Different Mode of Active Inference

ADHD: A Different Dance of Attention and Energy

Dyslexia: A Different Lens of Detail vs Holistic Thinking

Synesthesia: Blended Senses and Enhanced Experiences

A Personal Perspective

The Question of 'But Are You Really?'

Some Additional Reflections

**CHAPTER 26**  **Navigating Religion, Spirituality, and Pitfalls**  253

Opportunity and Boundaries for Religion in Society

Benefits of Religion and Spirituality

Horoscopes

A Possible Lens: Active Inference and Free Energy

Conversations Across Meanings

**CHAPTER 27**  **Training Our Bodies for Active Inference**  265

Adaptive Updating Through Embodiment

A Model for Embodied Updating

Adaptation Over Time

**CHAPTER 28**  **Reframing "Truth": Understanding the Limitations of Our Perceptions**  275

CHAPTER 29    **The Science Within: Believing in Our Built-In Capabilities**    285

Robust Science

Our Physiology: A Remarkable Support System

Beyond Awareness: Believing in Our Inherent Processes

CHAPTER 30    **Hopeful and Noteworthy Applications**    293

Verses: Eco-friendly Artificial Intelligence

Cortical Labs: Biological Computing Made Accessible

As a Tool for Governance

Computational Psychiatry: Personalized Paths to Healing

CHAPTER 31    **Getting Creative: Carrying This Forward**    301

CHAPTER 32    **Self-Knowing: Literally**    319

CHAPTER 33    **Opportunity for Action**    327

Inaction is a Form of Action

Know Your Strengths

Foster Community

Closing

**Keeping This Book Accessible**    332

**Acknowledgments**    333

**References**    336

**Index**    396

# FINDING HARMONY AMONGST UNCERTAINTY

# FINDING HARMONY AMONGST UNCERTAINTY

We have a lot going on as humans. We're striving to stay orchestrated in an unpredictable and continuously changing world, all while juggling the intricate demands of our mental, physical, and environmental needs. To make things even more confusing, these demands are often presented amidst conflicting cues and signals. This onslaught of information and competing priorities can feel like trying to perform a symphony without a conductor—a jumble of competing parts that overwhelms rather than enriches.

However, by understanding larger principles and how they interconnect (including our experiences, social sciences, human physiology, down to theoretical physics and back up again) we gain the tools to cultivate an inner conductor that orchestrates greater clarity and balance in our lives. When we devote time to integrating these understandings, cohesion grows within us, self-awareness strengthens, deeper resilience takes root, and uncertainty becomes something we can meet with expanded confidence.

This process isn't about forcing control but aligning with our natural rhythms, much like a symphony finding harmony under a skilled conductor. Ultimately, musicians are limited by the capacities of the human body. Yet through collaboration with the conductor they create beautiful experiences for

**ANNOTATION**

We use environment to mean the things around us, which is a wider sense than its meaning in climate or nature.

themselves and the audience. These principles function like this conductor, providing some direction across the interaction of mind, body, and environment within ourselves and between each other to find coherence, even amidst life's uncertainties and change.

Like many skilled musicians, this understanding can be embodied: supported both by mental cognition and bodily action and sensation. Similar to a musician mastering their instrument through continued practice, cultivating embodied understanding and behaviors requires integrating these principles into the flow of daily life. In this sense, our willingness to accept our human limitations and practice becomes a fundamental part of the bridge from basic understanding to experiential harmony.

This book seeks to support the learning and practice of increasing life's harmonies. This is done by simplifying complex theoretical biology, physics principles, and intricate formulas into relatable narratives addressing relevant concepts for today's challenges. It anchors itself in active inference and free energy—two foundational principles in science that are often shrouded by advanced formulas and abstract applications. Our hope is to bridge science and lived experience so it supports our life journeys, creating an accessible and embodied exploration that turns the overwhelming noise of modern life into the harmony of a beautiful life symphony.

"When we devote time to integrating these understandings, cohesion grows within us, self-awareness strengthens, deeper resilience takes root, and uncertainty becomes something we can meet with expanded confidence.

Navigating Uncertainty

# A STORY: HOW THIS BOOK CAME INTO BEING

CHAPTER 2

# A STORY: HOW THIS BOOK CAME INTO BEING

Sometimes, the smallest moments set the stage for the biggest transformations. For me, it began on an ordinary afternoon during the height of the COVID-19 pandemic. Life had ground to a halt, leaving us isolated in our homes, longing for connection and clarity. I lived in the Bay Area then, and on that day, stepping away from the relentless cycle of online meetings to sit outside under the California sun, I realized how profoundly unfulfilled I felt.

On paper, I'd done everything right. My résumé told a story of success: advanced degrees, a career in Big Tech, a life of achievement. But under these accolades, my life felt very different. I was chasing a version of success that left me constantly striving, anxious, and drained.

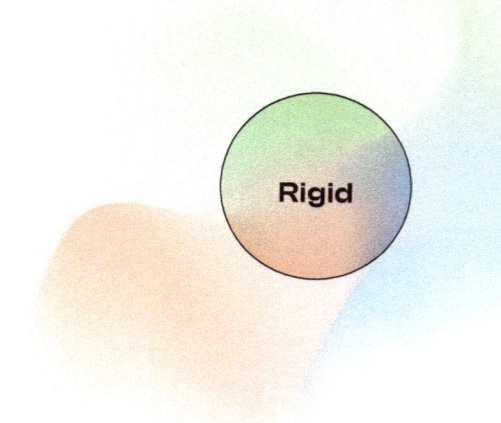

The cracks in the foundation had been there for a long time, but that day I felt them more acutely than ever. My career, social life, and sense of purpose—all of it felt like a house of cards on the verge of collapse again. I kept asking myself, *"How did I get back here? Why do the same patterns keep repeating?"*

And then it hit me: there was something fundamentally wrong in the system I had built.

For years, I'd been running on a glitchy autopilot, driven by external measures of success and validation. I chased a career I didn't love, built relationships that lacked depth, and ignored the quiet dissatisfaction simmering beneath the surface. I was trapped in cycles of avoidance, numbing myself in socially acceptable ways to cope with the dissonance.

That day, I realized the goals I pursued were misaligned, and the tools I relied on were equally off the mark. My efforts were fueled by ego and a deep hunger for external validation as a patch for unresolved tension within. I had reached the car I'd been chasing all my life, only to realize it was empty.

But that moment of reckoning became a turning point. A deeper realization took hold: to find what I needed to stop chasing the empty car and to help others do the same. I wanted to build a life guided by intentions and means that truly served me, rooted in a deeper understanding of how to make the most of my single experience of life, and to help others seeking to make their own shifts do the same.

I left Big Tech behind, packed my life into storage, and embarked on a transformative journey. I spent months living in an RV, immersed in nature and introspection. I devoured books, attended courses, and worked with a coach and therapist to resolve internal dissonance. Science had always been my compass, and I dove deeply into research, seeking a model that could integrate everything I was learning to share with others.

And then I found it.

At a conference brimming with innovators and thought leaders, I encountered paradigms that shifted everything: the active inference cycle and free energy principle. It was a scientific model, but it was also a deeply human one. It revealed how we navigate uncertainty, adapt to change, and have the opportunity to shift our lived reality.

As I learned more about active inference, its principles began to transform my daily life. Eventually, I moved on from simply learning new ideas to living them. It reshaped how I ate, moved, connected, and rested. It challenged me to rethink old patterns and make space for new possibilities.

The tools I had found aimed less at perfection or control and more at harmony and adaptation, helping me embrace both the messiness and beauty of being human. Being human is more than being something we "master." It's a constant unfolding, a dance between uncertainty and clarity. Similarly, this book isn't a guide to perfection, but rather an invitation to explore, adapt, and grow.

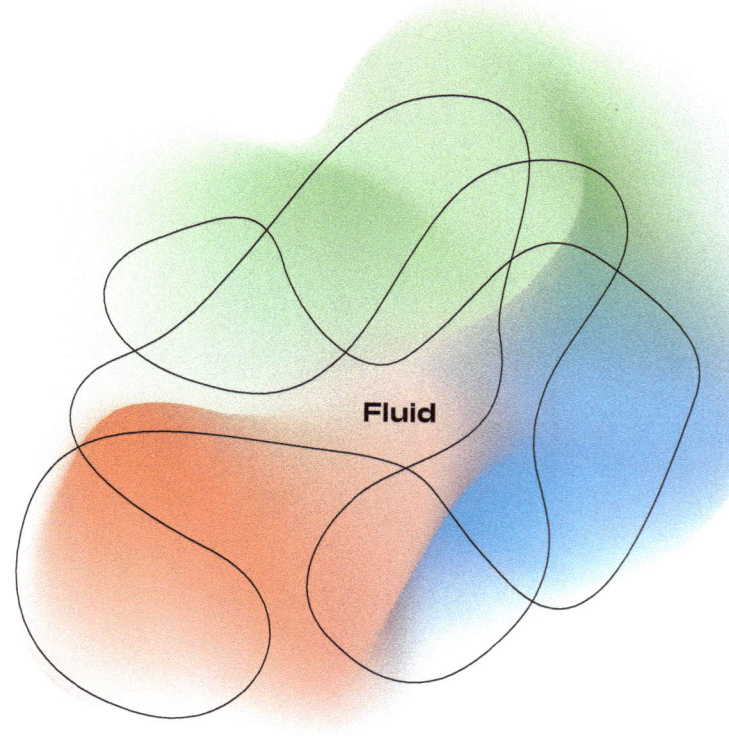

Fluid

This book grew from that journey. As I continued learning and integrating active inference, I realized I had been stuck in ineffective cycles for years, despite dedicating significant resources to breaking free. I had searched desperately for a way off the wheel, and I kept seeing others in the same position. Many were lacking either the opportunity or the know-how to do so. The skillset I had built in Big Tech and academia, combined with my transformative experience, now served a new purpose: to create this book.

My hope is that as you journey through these pages, you'll discover not just the science behind active inference and free energy but its potential to shift your own life. You might be navigating big changes, or simply seeking ways to adapt to the increasing pace of life. This collection of science-based concepts offers a powerful resource to support your goals. No single path looks the same, but in learning to navigate uncertainty with clarity, we gain increased freedom to shape our own unfolding.

# UNDERSTANDING OURSELVES THROUGH PHYSICS

# UNDERSTANDING OURSELVES THROUGH PHYSICS

This book combines robust scientific principles with the everyday experience of being human. In writing it, we seek to simplify and contextualize complex ideas, making them accessible and relevant in our daily lives, especially as we navigate growing levels of uncertainty. However, in simplifying, some specifics such as technical nuances, precise definitions, or subtle distinctions may be lost. This is a phenomenon we encounter in daily life. Even conversations with those closest to us lose depth and meaning when translated from feelings and ideas into words.

Despite this limitation, we strive to present the clearest and most accurate understanding of this science available today. Rather than postponing indefinitely without these valuable scientific insights, this book offers a timely snapshot of current knowledge, equipping us with a highly effective toolset to navigate our lives more effectively today.

## The Role of Science in Clarity

Navigating the fears, anxiety, and uncertainty of contemporary life is a pressing concern, with many well-intended narratives and ideas available to guide us. However, some stray significantly from useful or scientifically supported insights. This often complicates our ability to identify and apply the most valuable information amidst a sea of conflicting and occasionally misleading messages. This

**Active inference cycle**

**Illustrated and simplified active inference cycle**

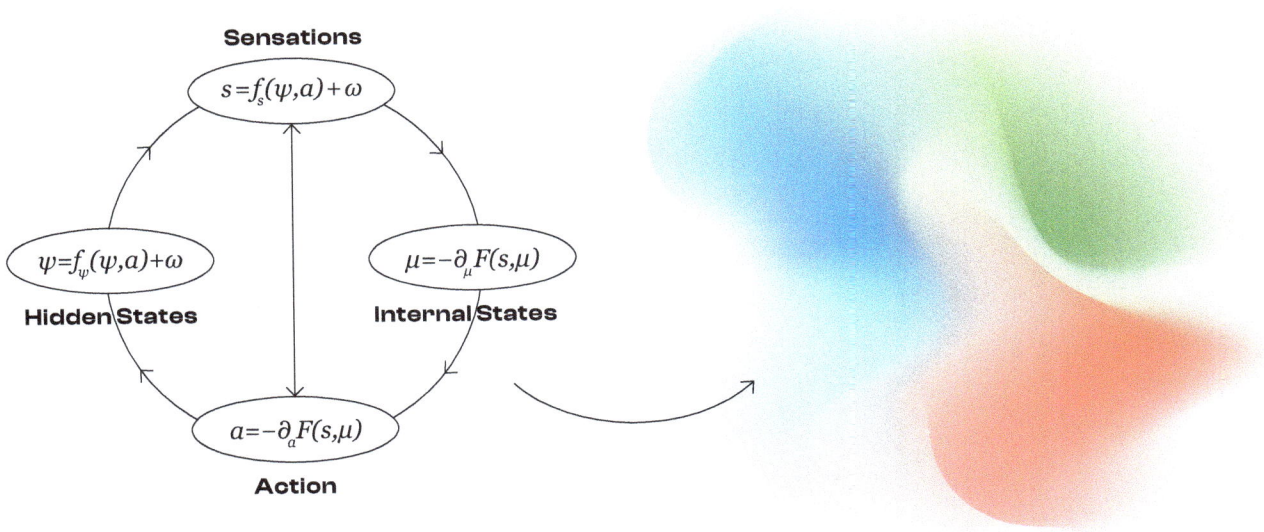

Sensations

$$s = f_s(\psi, a) + \omega$$

$$\psi = f_\psi(\psi, a) + \omega$$

**Hidden States**

$$\mu = -\partial_\mu F(s, \mu)$$

**Internal States**

$$a = -\partial_a F(s, \mu)$$

**Action**

book aims to offer clear guidance, deeply rooted in robust science, without limiting itself to scientific boundaries alone.

There are different ways sciences can achieve "robustness." One is through foundational mathematics paired with the ability to test hypotheses, an approach on which active inference is based. Active inference and free energy also achieve robust science by being supported by multiple peer-reviewed publications, with few contradictory findings, and reinforced by overwhelming additive results across diverse fields of study. For example, a quick Google Scholar search reveals over 16,400 results for "Free Energy Principle," highlighting its substantial and rigorous area of research.

However, we also recognize the inherent limitations of science. Just as various traditions interpret the same foundational text differently, there are multiple schools of thought within the Free Energy Principle. This book takes a high-level view, favoring simple explanations while acknowledging that deeper interpretations exist and that such diversity strengthens the field over time.

## From Uncertainty to Understanding

To illuminate our path forward of active inference and free energy specifically, we can also consider its origins. The free energy principle and active inference cycle were first articulated and developed mathematically by neuroscientist Karl Friston primarily during the early 2000s. If you haven't encountered free energy or active inference yet, that's fairly typical. It typically takes between 40 and 100 years for new theoretical concepts to become part of the common vernacular, and at the time of writing, we're still relatively early.

We are fortunate to be living at a moment in history when Karl Friston and his colleagues have already laid down such a vast body of work. Karl himself is a British neuroscientist trained in both medicine and mathematics, whose career has been defined by bridging rigorous theory with practical applications. Known for his meticulous methods and cross-disciplinary approach, he integrates insights from physics, neuroscience, and information theory to explain how living systems sustain themselves. He's often regarded as the "grandfather" of the Free Energy Principle and one of the most cited and influential scientists today. His work has set a stage that allows the rest of us to explore, apply, and communicate these ideas with far greater clarity and accessibility.

In the case of this book specifically, Karl Friston generously contributed directly. He kindly advised, reviewed, and even wrote the foreword, all as part of a conscious effort to ensure the science and communication are as effective and accurate as possible.

## Active Inference and Free Energy

Active inference and free energy offer insightful models for understanding how living systems manage uncertainty. Fundamentally, these theories demonstrate that we, and other life, manage change and unknowns by aligning internal expectations (what we think will happen) with external reality (what actually happens). We then work to actively reduce the discrepancy between the two.

To illustrate, consider a situation where uncertainty triggers anxiety, such as navigating the high levels of chaos in contemporary life. Active inference suggests that consciously holding various possible outcomes in mind, without prematurely settling on one, helps us remain balanced and responsive. Emotionally, this means being able to experience and process feelings like anxiety or confusion without immediately seeking resolution. In terms of free energy, maintaining open rather than fixating prematurely reduces internal prediction errors, essentially lowering our likelihood of incorrect predictions about the world. This lowers our free energy, the felt tension in our lives, and enhances our overall sense of stability and connection over time.

**ANNOTATION**

We also want to acknowledge that, given the realistic limitations of self-funded research required for the complex systems thinking of this book, we used AI as a supporting tool. Human authors and experts were always the originators, reviewers, and final approvers of the work.

**ANNOTATION**

By holding space for these emotions, we also remain better connected with others, as openness to uncertainty often fosters empathy and mutual understanding.

# CHARTING YOUR COURSE

# CHAPTER 4
# CHARTING YOUR COURSE

Reflecting back on Chapter 1, we saw how fundamental principles of active inference and free energy equip us to cultivate an internal conductor, facilitating greater clarity and balance in our lives. Continuing with this metaphor, we instinctively appreciate that musicians and conductors rely heavily on dedicated practice and rehearsal (both individually and together) to achieve harmonious performances. Learning involves acquiring knowledge or skills through study or observation, often remaining an intellectual process that may not directly influence actions, such as mastering the ability to read musical notation.

Embodying, however, goes beyond understanding to deeply integrate knowledge into daily actions and habits, aligning thought, behavior, and being, such as playing a deeply expressive and mastered piece. Similarly, the activities in this book serve as your rehearsal space, providing opportunities to integrate these scientific principles into your daily life. The activities are adaptable to your specific needs, encouraging you to tailor them to your journey and what resonates for you. Much like musicians dedicate time to practice, consider setting aside moments for your own personal practice, enriching your experience, and enhancing the benefits offered by this book. In turn, these opportunities not only resonate intuitively but, as integrations of mind, body, and environment, are fundamentally supported by active inference.

**ANNOTATION**

We want to pause to acknowledge here that while many people benefit from learning how to read music, not everyone requires this skill to enjoy or excel in playing an instrument, and the same goes for learning this science. It's a useful path for many, but not necessarily required.

This book is designed with a gentle order in mind. However, if you feel called to explore it in a different sequence, the content is crafted to stand alone, allowing you to follow your own path. As you make your way through this book, we encourage you to also thoughtfully engage with the material. By consciously interacting with these materials, you nurture the integration of mind and body, fostering a deeper, more intuitive understanding of yourself and your environment. For example, what resonates with you? What might you want to revisit? For things that don't resonate, consider why that might be. We also welcome your feedback to help shape future iterations of this work. You can share your thoughts here: www.actincycle.com/feedback

# THE ACTIVE INFERENCE CYCLE: OUR LIVED EXPERIENCE

CHAPTER 5

# THE ACTIVE INFERENCE CYCLE: OUR LIVED EXPERIENCE

Now that we've laid the groundwork for this book, we can begin to explore the core concepts more deeply. First among them is the active inference cycle—a foundational thread that runs throughout the book. The active inference cycle is a scientific concept used to describe how our brains continuously sense and interpret the world around us, test those interpretations through our actions, and then adjust based on the feedback we receive from our environment, body, and mind.

In short, active inference is the foundational process underlying our existence. Examples of this include responding to unexpected events like a sudden loud noise, which triggers immediate awareness and recalibration to restore calm; anticipating audience reactions during presentations, allowing us to modify our communication dynamically; intuitively stepping back from a passing car, seamlessly integrating sensory input with protective action; and sensing subtle shifts in your own energy during meditation, gently adjusting your breathing or posture to maintain harmony and focus.

Much like the felt complexity of life, active inference spans multiple realms, including physics. What makes this particularly useful and fascinating is that the human brain is highly adept at making predictions and inferences rooted in physics. We rely on these predictions constantly, whether catching a ball, navigating traffic, or intuitively anticipating the conclusion of a captivating novel. Just as we take the time to learn how to throw a ball, drive a car, or read, we can also pause to understand these principles. In doing so, we uncover the fundamental physics behind our experiences, offering an opportunity to harness our powerful innate ability to predict physical processes as a resource for navigating life's challenges. Much like mastering any skill, once we begin to learn these concepts, we can begin to practice and integrate this knowledge more deeply into our daily lives.

We can start by breaking it down into smaller parts. Throughout the book, you'll hear us consistently refer to the mind, body, and environment. Everything we perceive in life unfolds across these interconnected realms, which continuously shift in prominence and influence.

**THE MIND:** This is where we make sense of the world, form expectations about what might happen, and align our goals with our actions. Our intentions guide our choices and help us manage the uncertainty that surrounds us. By staying open to new information and adjusting our thoughts and actions, the mind helps us adapt and move through life's challenges.

**THE BODY:** The body is a foundation, housing the mind, acting as a sensory organ, and enabling us to interact with the world. Through action and sensation, it bridges the gap between the mind and environment, providing continuous feedback. From subtle interoceptive signals to the tangible effects of our movements,

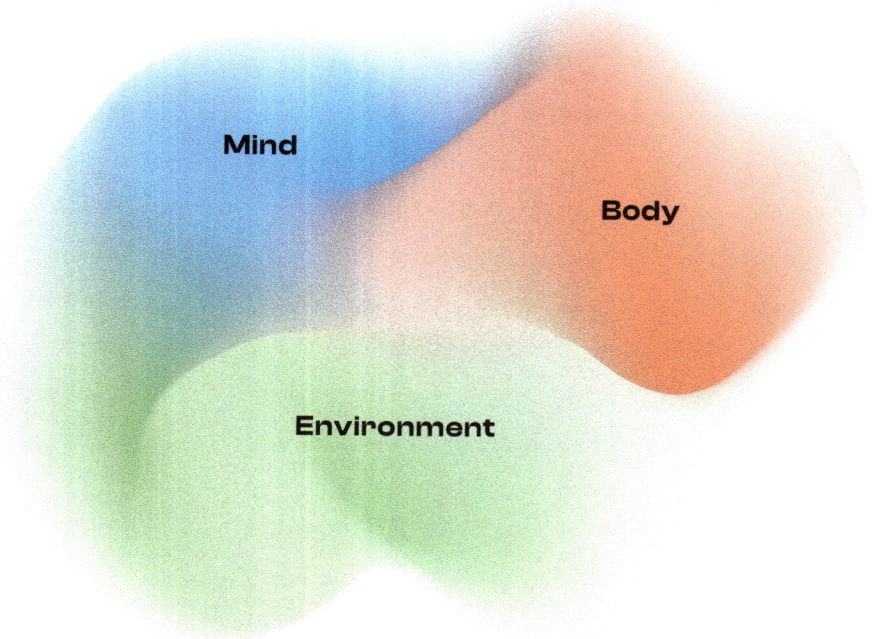

Mind

Body

Environment

they shape our ongoing predictions and refine our responses. Without the body, our existence ceases; it grants us the gifts of intuition while ultimately enabling connection with the world around us.

**THE ENVIRONMENT:** External factors, including our relationships, culture, and surroundings, act as both the source of inputs and the context for our actions. The environment around us profoundly influences both our mind and body, shaping how we perceive, react, and adapt. For example, environments characterized by rapid change, such as shifting workplaces or evolving social dynamics, require us to process uncertainty and develop new behaviors to adapt—a reality many of us face in our daily lives.

These three components are deeply interconnected, constantly influencing and shaping one another. One such example is the gut-brain axis: bodily states, such as digestion, can affect mental clarity, while mental clarity can directly impact gut health. Moreover, the environment provides the gut microbes essential to the gut-brain axis, further highlighting the dynamic interplay between these spaces. The boundaries between mind, body, and environment are porous, with each seamlessly flowing into and shaping the others.

At times, one domain may dominate another. During periods of deep thought or reflection, the mind may take center stage. When we're physically ill or focused on movement, the body's role becomes more pronounced. In social or cultural contexts, the environment often exerts the greatest influence. Yet to truly align with the flow of life, it's necessary to reflect and foster all three spaces and their interplay.

This book is crafted to honor this balance. We've designed it to explore the mind, body, and environment in equal measure, acknowledging their shifts and interplay. By doing so, we aim to reduce bias and foster a more integrated understanding of self-knowing, adaptability, and resilience.

As we move through the chapters ahead, we'll explore each of these spaces, exploring how they contribute to the active inference cycle. However, these principles aren't just theoretical; they're a lived experience, shaping how we grow, connect, and find harmony in the flow of our lives.

"The boundaries between mind, body, and environment are porous, with each seamlessly flowing into and shaping the others. To truly align with the flow of life, it's crucial to reflect and foster all three spaces and their interplay.

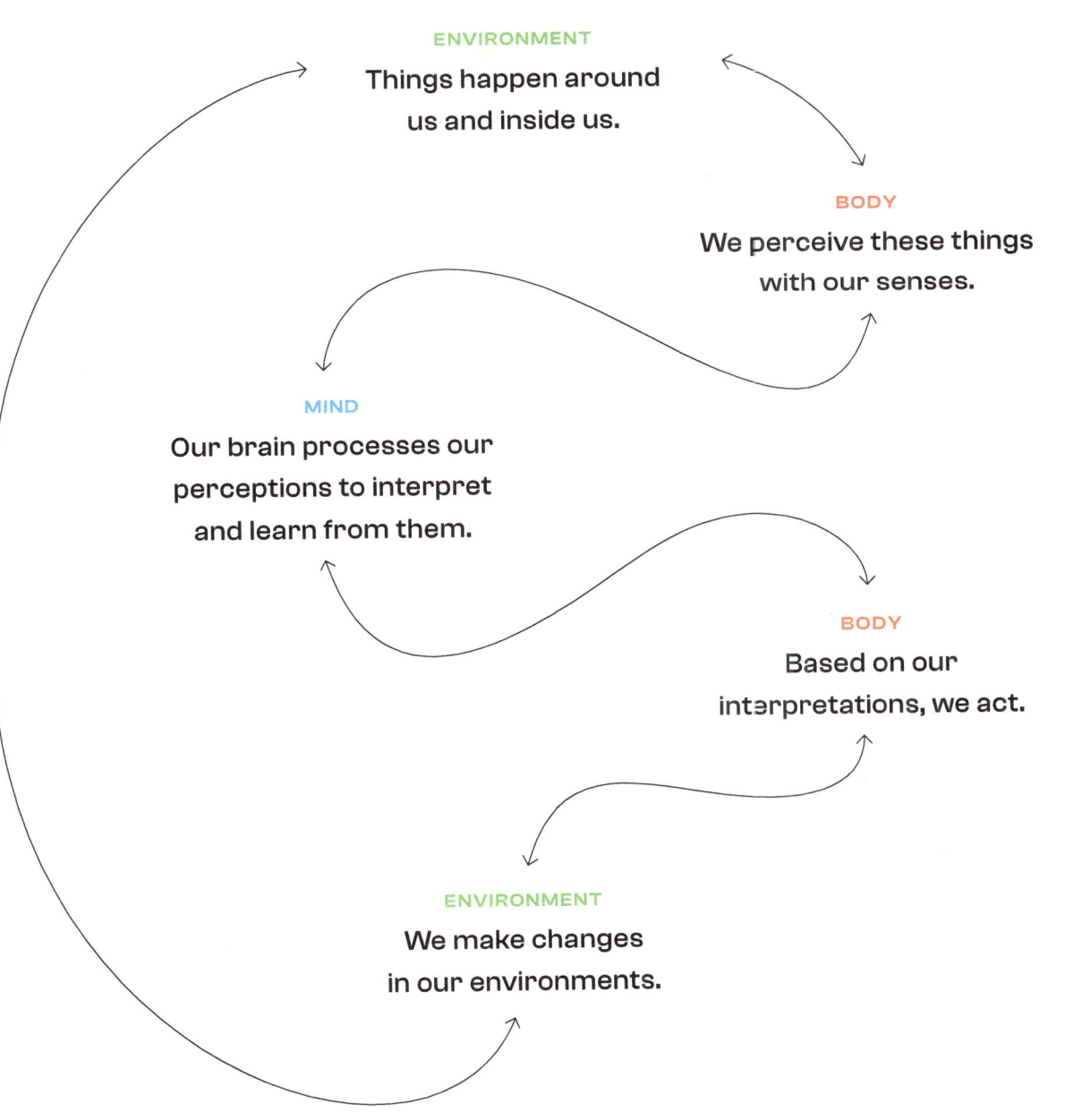

**ENVIRONMENT**
Things happen around
us and inside us.

**BODY**
We perceive these things
with our senses.

**MIND**
Our brain processes our
perceptions to interpret
and learn from them.

**BODY**
Based on our
interpretations, we act.

**ENVIRONMENT**
We make changes
in our environments.

# THE FREE ENERGY PRINCIPLE: SEEKING HARMONY & DISSONANCE

CHAPTER 6

# THE FREE ENERGY PRINCIPLE: SEEKING HARMONY & DISSONANCE

If active inference represents the flow of everything (including your life), then the free energy principle is the driving factor behind that flow. The free energy principle suggests that every action, thought, and feeling is part of a larger effort to reduce surprise and uncertainty in your life. Its function is to align your internal expectations with the world around you. When things don't go as expected, you feel discomfort or tension (free energy), prompting you to act, reflect, or adapt to bring yourself back into balance. It is the deeper underpinning that drives you to create harmony between what you expect and what you experience.

Interestingly, the free energy principle runs counter to what we often think we want. Instead of wanting more free energy in our lives, we generally seek less. We want to avoid excessively wasted energy, large unexpected surprises, or continued themes of dissonance—a persistent sensation that feels out of sync or carries risk to our overall stability. However, some surprise plays a foundational role in how we make sense of the world and make decisions. It's about finding a balance. For example, if we locked ourselves in an empty room, we might initially appreciate the quiet sanctuary of the space, but after a while, we would get quite bored. Rather, an activity like a walk in a park would more readily balance our expectations while still allowing for stimulation.

Overall, we make sense of things by finding explanations that render our observations the least uncertain while also acting in ways that maximize a specific kind of surprise. For example, consider unwrapping a birthday present. The act itself introduces temporary uncertainty but ultimately satiates our curiosity by resolving what lies inside the gift. In this way, we actively seek uncertainty-resolving experiences as part of minimizing uncertainty in the long term.

Just as the active inference cycle helps explain our ongoing interactions with the world, we can explore how the free energy principle manifests in our daily lives through the interconnected domains of mind, body, and environment.

**MIND:** We might experience free energy mentally when we encounter a moment of uncertainty when trying to recall a key word during a conversation. The inability to retrieve the word creates tension, as it feels like something is missing or out of place. This mismatch, or free energy, drives the mind to search for a resolution by scanning memories or even feeling compelled to ask someone else. Once the word is remembered or the mind shifts focus away from the gap, the tension dissipates, restoring a sense of mental balance and flow.

**BODY:** One way we physically experience free energy is through our sense of temperature, which helps maintain a thermal internal balance essential for survival. When we detect a drop in temperature, our bodies signal the need for warmth to prevent disruptions to vital processes. Feeling cold prompts the action of putting on a sweater, which restores balance by aligning the external environment with the body's internal requirements. This simple act reduces free energy by resolving the mismatch between the body's need for warmth and the external coolness in the room, ensuring comfort and stability.

**ENVIRONMENT:** Perhaps one way you're already familiar with free energy in your environment is how working in a quiet, harmonious space that aligns with your expectations (low free energy) can lower stress and help you focus. In contrast, a noisy, chaotic environment filled with unexpected interruptions (high free energy) can require constant adaptation, potentially breaking your flow and increasing strain.

I've come to appreciate the deep sense of joy and relief that comes from engaging in activities and experiences that resolve uncertainty in a manageable way, even when they involve temporary discomfort. For example, noticing and sitting with subtle, persistent dissonance within myself during an extended period of silence, and then experiencing a deep sense of relief as it gradually resolves. Or navigating a challenging set of emotions in order to have a meaningful conversation with my partner. Even listening to music that is mostly calming but weaves in moments of dissonance that eventually resolve can be quite satisfying. Likewise, attending a large event with colleagues and friends, where I may not know exactly what I'll experience but trust that I'll have opportunities to connect meaningfully and gain valuable insights from the experience.

Taking this concept a level deeper, the free energy principle suggests that the undercurrent of our lives isn't centered around optimizing for external markers like wealth or status, as society often emphasizes, but rather on reducing uncertainty and resolving internal instability. This involves deep self-awareness, learning to understand more of our surroundings, and aligning our actions to maintain balance and resilience. Engaging with the world for intrinsic value (such as seeking knowledge and understanding) resonates more with this principle than chasing external rewards, such as promotions or superficial beauty standards. When our focus aligns with the underlying physics of experience and the drive to resolve dissonance, a more fulfilling and sustainable way of living comes into view. In essence, we support what the free energy principle is naturally guiding us toward: seeking knowledge and understanding while reducing dissonance and tension, allowing us greater ease, adaptability, and resilience in an unpredictable world.

**Navigating Uncertainty**

# Active Inference Cycle Experiencing
## Normative Levels of Uncertainty

# HOMEOSTASIS: A FOUNDATION OF LIFE

# HOMEOSTASIS: A FOUNDATION OF LIFE

As we explore the active inference process, it naturally highlights the essential role our bodies play in helping us stay balanced and well. Yet in many modern societies, we often relate to our bodies with guilt, pressure to change, or neglect. We forget or sideline that our bodies, far from passive vessels, take an active and supportive role in the rhythms of daily life. One of those supportive processes is homeostasis, a quiet, behind-the-scenes process that keeps us alive and steady. It allows our bodies and minds to recalibrate continuously, helping us function smoothly even in the face of daily disruptions. Without it, balance would break down, leading to dysfunction and eventually the end of life.

Homeostasis is also the bedrock of resilience and sustainability. It ensures that the body, and the embodied brain, can adjust, reorganize, and recover when challenged. This adaptive ability, often referred to as allostasis, is what enables us to move through difficulty and return to center. In this way, homeostasis works to keep things steady, realigning us back to balance again and again, even as the movement of life continues around us.

Imagine you're at a social gathering. Basic needs like air and fluids are covered, but internal balance, such as emotional calm or mental clarity, might need more attention. So, you step outside for fresh air, away from the crowd and noise.

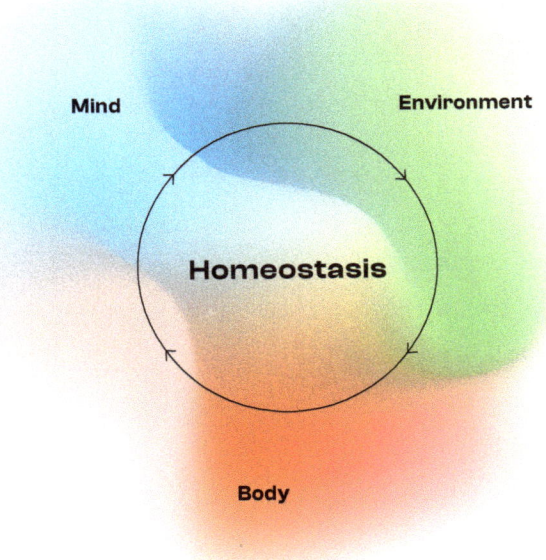

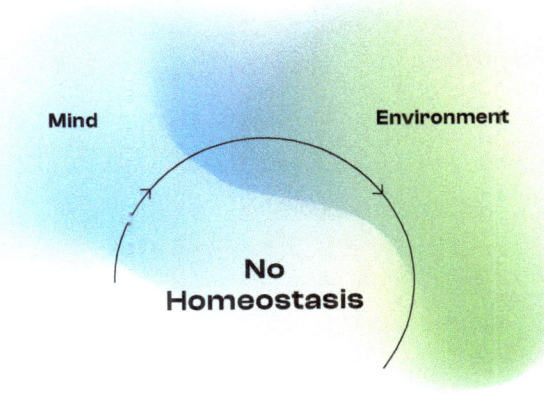

That moment of pause helps your system reset, and you return more grounded and present. That's homeostasis in action. As we practice tuning into those needs and responding with attunement, we build the muscle of resilience. With time, we learn how to meet life with more ease, adaptability, and poise.

All life engages in homeostasis in some form. For humans, we can continue to organize these needs into the categories of mind, body, and environment, each deeply shaped by our physiology.

**MIND:** Our cognitive well-being depends on a delicate dance between predictability and novelty. Too much uncertainty can overwhelm us; too much sameness can stunt growth. We need stability to feel safe and uncertainty for the opportunity to learn. This intricate balance is one of the tricky yet essential dynamics this book aims to help foster.

**BODY:** Our physical systems rely on constants like temperature regulation, nourishment, oxygen, and sleep. These needs might seem obvious, but they're often sidelined in fast-paced cultures. When ignored, they disrupt the broader cycle of active inference, limiting our potential. However, it's remarkable that our bodies handle so much of this automatically. They offer us this regulation as a daily resource—a quiet contribution towards steadiness and adaptability.

**ENVIRONMENT:** We need safe environments that support both our mind and body. This includes essentials such as water free of toxins or harmful pathogens, as well as physical spaces that protect us from violence, abuse, or chronic fear. Our environments also need to meet our psychological needs, providing social bonds rooted in trust, a sense of belonging within a group or culture, and mutual support in times of need. These elements help create the stability required for resilience and the ability to sustain ourselves over time.

**EMOTIONAL NEEDS:** Emotional needs are deeply intertwined with the mind, body, and environment. To sidestep any academic debates in this category, we treat emotional needs as a combination of these interplays. We need emotional safety and the ability to express ourselves without fear of harm or rejection.

Stress regulation is also critical, whether through mental or physical coping mechanisms or supportive environments. For many, emotional needs extend to a sense of purpose, meaning, or inner harmony.

All of these layers connect back to one of this book's central themes: reducing free energy in our lives. When our needs are met across these domains, we generate less inner conflict and more inner coherence. That coherence supports our ability to maintain homeostasis, stay steady in uncertainty, and ride the waves of life instead of fighting the tide.

## SPARK

- Take a moment to consider your own balance of homeostasis. What's one area of your life (mind, body, or environment) that currently feels strong or supportive? What practices or conditions help maintain that strength, and how might you continue to honor and preserve it?

- Now consider an area that feels a little off balance, but within reach of realignment. What small, realistic shift could support greater alignment there? What supportive, actionable step could you take this week to begin that shift?

- Do either of these prompts spark ideas for how you might support others in nurturing their own homeostasis, balance, and adaptability?

# SELF-KNOWING: TURNING UP THE LIGHT

# SELF-KNOWING: TURNING UP THE LIGHT

Just as homeostasis quietly steadies your body's internal rhythms, self-knowing gently stabilizes the rhythm of your inner perceptions. Both are ongoing, adaptive processes that continuously recalibrate the interconnection between physical sensations and mental experiences. More specifically, self-knowing is an ongoing process of making sense of how your inner world and the outer world interact. It's a bit like slowly turning up a dimmer switch—at first, things are hazy and undefined, but as the light grows, patterns begin to emerge. You start to see how your thoughts, feelings, and actions are woven together, influencing both your perception and your choices.

Grounded in the science of active inference, self-knowing means constantly updating your understanding based on new information. This could include your best guesses about yourself and the world or your community. As the self-knowing light brightens, your ability to make effective and thoughtful choices grows, revealing opportunities that were once hidden in the shadows. This clarity allows you to navigate both your inner and outer worlds with greater ease and resilience.

## Deepening Self-Knowledge Through Action

Unlike a single adjustment, self-knowing can continue to deepen over time, uncovering striking insights as you move through the world. However, self-knowing is also greatly supported by enactivism: an interdependent interaction of mind, body, and environment. While reading this book is a meaningful first step, deepening your self-knowing will require incorporating actions as well. These actions are opportunities for meaning-making and processes through which we learn, understand, and shape both ourselves and our environment. Through these interactions, self-knowing becomes an iterative process and it evolves as we engage with the world and refine our understanding of it.

For example, engaging in a conversation about differing opinions, such as discussing climate change with a friend who holds a skeptical stance, can serve as a powerful practice for self-knowing. Perhaps their view is rooted in limited precision of the science, different hopes for the direction our world could take, or personal experiences. Regardless of the reason, their differing perspective might initially trigger frustration or fear, but exploring their point of view can reveal deeper layers of understanding. These interactions can be challenging as hidden assumptions or biases surface, yet they push us to refine our beliefs and broaden perspective. With practice, the emotional responses such conversations provoke become easier to manage, revealing what drives our reactions and how we communicate under tension. This process shows how self-knowing is shaped through real-time interaction between mind, body, and environment. You're not just exchanging ideas; you're participating in something that reshapes you as it unfolds. Such exchanges illustrate how self-knowing evolves through active participation in the world.

We've included several options to practice and foster greater self-knowing throughout the book. Perhaps one or two will resonate with you and inspire you to take a moment to set the intention to explore them. When these connections are brought to light through understanding and action, self-knowing equips you to move through life with greater ease, resilience, and purpose.

## The Art of Balancing Precision

Active inference and self-knowing offer many perspectives on how we can deepen our understanding of the world and selves. At the most basic level, we can remember these processes involve the body sensing what's happening in the world through inputs like sight, sound, and interoceptive signals, and the mind supporting the creation of meaning from these sensations. For example, if you were to look up from this book right now, your visual system automatically processes sensory details, like edges and corners. Building on these basics, higher-level perceptions emerge as your mind supports integrating these details into meaningful interpretations—like recognizing a chair or a tree in front of you.

This progression from raw sensory input to understanding extends far beyond simple examples. It contributes to shaping nearly everything we interpret and understand about the world. It applies to adapting to new technology, reflecting on personal growth, navigating complex relationships, or making sense of global events. Self-knowing plays a critical role here: the deeper your awareness of how your internal filters operate as they progress from raw sensory input to complex interpretations, the better you can avoid getting lost in emotional reactivity, improve decision-making, and deepen interpersonal relationships.

One of the key filters we have the opportunity to shift through self-knowing is called precision—the level of confidence and focus we give our senses and interpretations. For example, consider experiencing intense emotions during periods of broader societal or personal uncertainty. Precision here means balancing how much focus you give to immediate emotional reactions versus the context of the larger situation. If your focus in your initial emotional response is too low, you might dismiss genuine feelings of anxiety or discomfort, leaving underlying issues unresolved. Conversely, placing too much focus in immediate emotions can amplify distress, obscuring the bigger picture and limiting your ability to respond constructively.

Self-knowing allows you to recognize and adjust the focus and confidence you assign to your perceptions. For example, in this case, it could mean recognizing personal patterns (such as knowing you sometimes lean toward anxiety) and intentionally creating space for your emotions without overly fixating on them. In utilizing this self-knowing, you don't suppress genuine feelings or allow them to dominate; instead, you consciously nurture a middle space. Ultimately, self-knowing allows you to deepen a more coherent and resilient understanding of both yourself and the world around you, particularly during periods of uncertainty or stress.

Another relevant example of how self-knowing and focus can help us cultivate greater clarity for ourselves is making sense of global events, akin to moving through a crowded marketplace bustling with political ideas.

## Unbalanced Precision
Ignoring signals or overfixating on them.

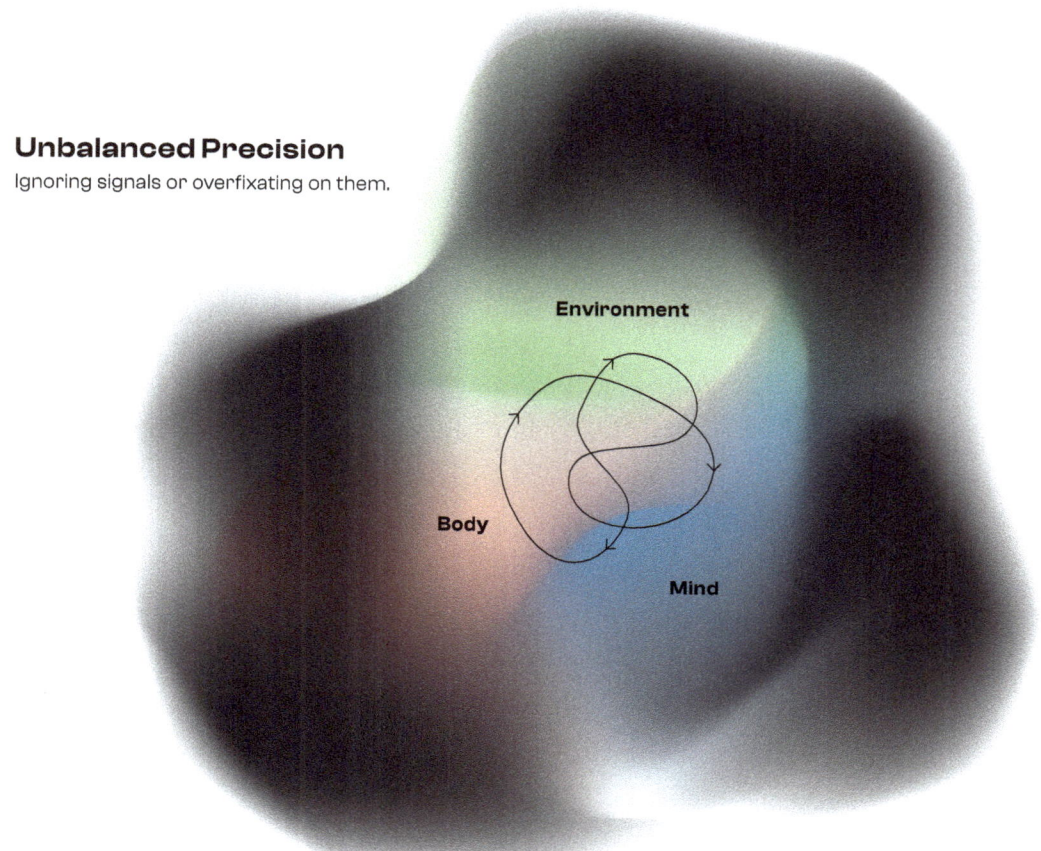

"Precision and focus" here means carefully assessing how much confidence you place in each source of information you encounter. If your trust is too low, you might become overwhelmed, unable to discern meaningful patterns amidst the noise. This would be as if you were standing in the market, skeptical of every vendor, buying nothing, and leaving empty-handed. Conversely, if your certainty is overly high, you might uncritically accept a narrow viewpoint as if you were

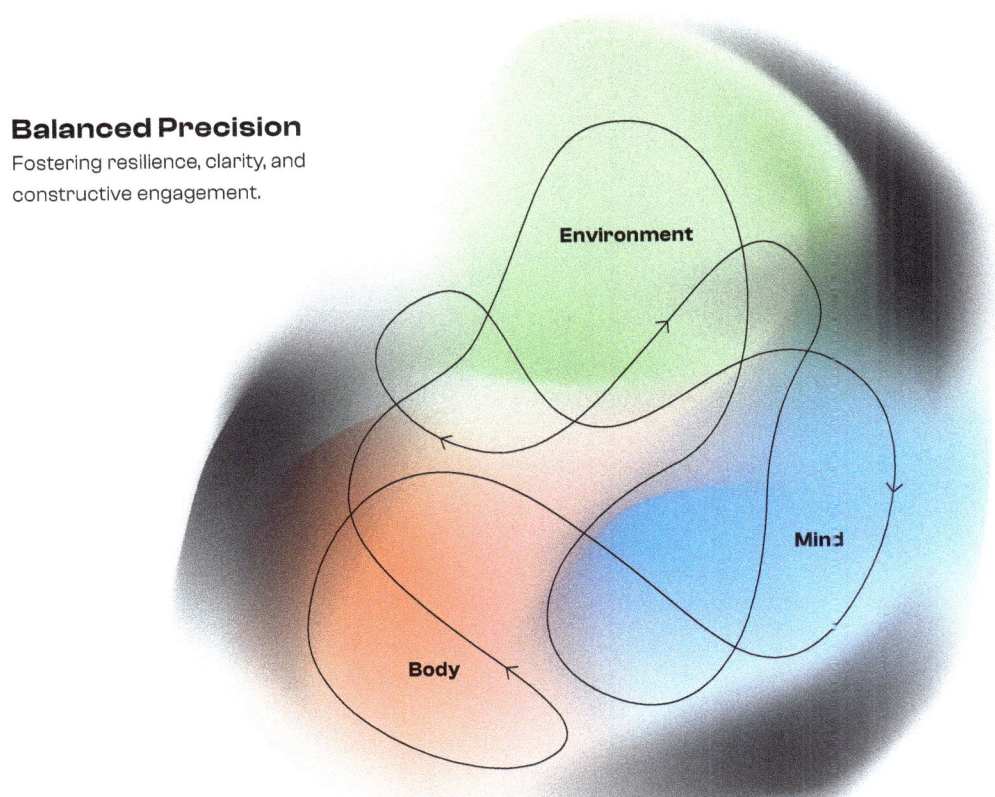

**Balanced Precision**
Fostering resilience, clarity, and
constructive engagement.

Environment

Mind

Body

to visit only one vendor whose ideas align exactly with your political affiliation.
This would result in possibly missing valuable perspectives or goods from other
stands.

Therefore, when we recognize our own political affiliations and biases as
part of our self-knowing, we gain the clarity needed to thoughtfully engage and
respond to diverse political perspectives. This awareness helps us identify how

each viewpoint might illuminate valuable truths or conceal important complexities. This thoughtful balance sharpens our understanding, strengthens our capacity to engage constructively with others in a complex political landscape, and steadies how we move through global events.

## Meditation: One Gateway to Balanced Precision

Awareness meditation, among many other practices, integrates into the process of self-knowing, providing a powerful tool to illuminate the patterns of your mind. Studies demonstrate that just eight weeks of awareness-based meditation can produce measurable changes in the brain.

Because our bodies play such a pivotal role in this process, we're limited in our ability to make the most effective use of our senses, cognition, and the confidence we place in them through thought alone. It's like trying to learn to ride a bike by only studying how to theoretically pedal. They are similarly limited. On this view, meditation and other practices enable one to develop the skill of relaxing the confidence placed in prior interpretations by tuning back into our senses. It's akin to learning how to ride a bike by practicing with training wheels. Eventually paving the way for us to do things like mastering riding a bike or, for some of us, performing mind-boggling feats like navigating intense mountain bike courses.

Meditation and similar activities also create a vital connection between our bodily sensations, mental interpretations, and the precision we assign to these experiences. For example, focusing on your breath during meditation can anchor

**ANNOTATION**

A notable study led by Hölzel and colleagues found that participants experienced increased gray matter density in the hippocampus, a region critical for learning and memory, and reduced gray matter in the amygdala, which plays a key role in stress and anxiety. These findings not only underscore the brain's remarkable neuroplasticity but also highlight the transformative potential of intentional practices such as meditation. Beyond these measurable shifts, practitioners reported significant improvements in their experiential quality of life, including enhanced emotional regulation, mental clarity, and a greater sense of calm.

you in the physical sensation of breathing, while also allowing mental patterns (like distractions or anxieties) to surface and become clearer. This heightened awareness recalibrates the confidence placed in entrenched beliefs or past assumptions that are often lurking and influencing our cycles below the surface. Realizing these subtle and often invisible cues early in the cycle helps to minimize their impact before they become integrated and potentially amplified later in the cycle. Refining this foundational process through meditation and similar practices cultivates a greater sense of balance and clarity while also sharpening the precision needed to handle complex decisions and relationships.

While many of us might not feel compelled to become highly committed meditation practitioners or extreme athletes like mountain bikers, we can still recognize their value as life skills. However, there's an important distinction: learning to ride a bike is optional, but the processes of sensing, interpreting, and assigning confidence to our perceptions are unavoidable. Just like learning a mode of transportation for yourself (such as walking or driving), cultivating practices to integrate the mind and body is a key foundation of our larger cycle. In a world that is shifting rapidly and asking more of us, practices like meditation can be an invaluable skill to nurture.

Because this is such a compelling and impactful set of ideas to understand, we want to offer one more illustration. Imagine your day-to-day experience like moving through a rugged landscape. Some paths are blocked by towering peaks of certainty and beliefs we've held so tightly that they obscure what might lie beyond. Other areas are carved into deep ravines, shaped by behaviors or thought patterns we've followed for so long they've become hard to escape.

Practices like meditation, or anything that helps us pause and reflect, can gently reshape this terrain. When we learn to ease the grip on our interpretations and those sharp convictions we've clung to, it can soften those high peaks, giving us a clearer view of what else might be possible. And when we become more aware of the behavioral grooves we're stuck in, we begin to climb out of those mental ravines. The key change in this case comes by slowly recognizing we might have more paths available to us than we thought.

Rather than trying to alter the landscape, we can loosen what has held us back. This allows us to move with more ease, a broader perspective, and greater adaptability. Over time, as we update our inner maps, the world around us opens up too. That's the quiet opportunity of these practices: they reveal what's been possible all along.

**ANNOTATION**

If meditation isn't your thing, that's okay. Some other ways to experience some of the same benefits include expressive practices like dance, art, and music; reflective activities such as journaling or certain forms of prayer; and grounding experiences in nature like walking, gardening, or simply spending time outdoors. Different approaches speak to different people—what matters is finding something that helps you tune in and connect more deeply with yourself.

## SPARK

- Consider experimenting with a new practice that invites self-knowing. It could be as simple as paying closer attention to your breath for one minute each day or taking a walk without your phone in nature. Or, if you feel drawn, perhaps it's trying something a bit deeper—like a new form of breathwork or attending a meditation retreat. And if you already have a practice that nourishes you and feels aligned, that works too.

- What's sitting under the surface as you explore this activity? What is it asking you to do? How can you realistically move forward with a small bit of action to further your self-knowing?

"When we recognize our own political affiliations and biases as part of our self-knowing, we gain the clarity needed to thoughtfully engage and respond to diverse political perspectives.

# ADJUSTING THE ZOOM & FOCUS OF OUR AWARENESS

# ADJUSTING THE ZOOM & FOCUS OF OUR AWARENESS

## A Simplified Explanation of Attention and Precision

We've just explored how shifting the focus and the precision (or confidence) we assign to things like emotions or political beliefs can contribute to deeper self-knowing. But this idea extends well beyond those examples. Precision and focus apply to all of our senses and interpretations. And when we combine this with where we place our attention (like adjusting the zoom on a camera) we gain a powerful way to shape our experience. We can play with both the zoom (attention) and the focus (precision) of our intentional awareness, and by doing so, we gain tools to more skillfully understand and navigate our world. These ideas, drawn from the science of active inference, show that perception is both receptive and actively shaped by us. Once you've zoomed in, the next step is to adjust the focus and precision. The sharper the focus, the more clarity we bring to what we're attending to—allowing us to more precisely understand and engage with what we've zoomed in and turned our attention to.

Let's consider another relatable example: being at a crowded cafe. There's a symphony of noise—the hum of conversation, the hiss of a coffee machine, the clink of cups. But if you're engrossed in a conversation with a friend, your brain zooms in and pulls its attention towards their voice. This act of selective attention ensures that, amidst a sea of information, you remain anchored to the present-moment priority. The focus (the precision) is the process of tuning out irrelevant sounds and strengthening your confidence in what you're hearing. It's a dance between two intertwined processes: selecting where to place your attention (zooming in on your friend's voice) and increasing the clarity and precision of what they're saying (by focusing on them and filtering out distractions).

On a neurological level, precision fine-tunes the brain's signal strength, akin to the conductor directing an orchestra—ensuring key signals are amplified while irrelevant ones are dampened. Known as "synaptic gain," this process dynamically balances neural excitation and inhibition, prioritizing the most relevant information for action and perception.

Many of these processes happen automatically and work in our favor. For example, consider saccadic suppression. As you're reading this book or moving through your day, your brain automatically blurs visual input each time your eyes shift. Your brain handles this suppression so effectively that unless you were already aware of saccadic suppression, you probably never realized your vision briefly stops processing detailed visual information each time your eyes shift position. Without this, the constant motion from your eye movements would likely overwhelm your visual system. This built-in mechanism prevents disorientation, allowing you to perceive a stable world even while your gaze is constantly in motion.

# Zoom

To select where you turn your attention.

**Navigating Uncertainty**

# Focus

To refine precision.

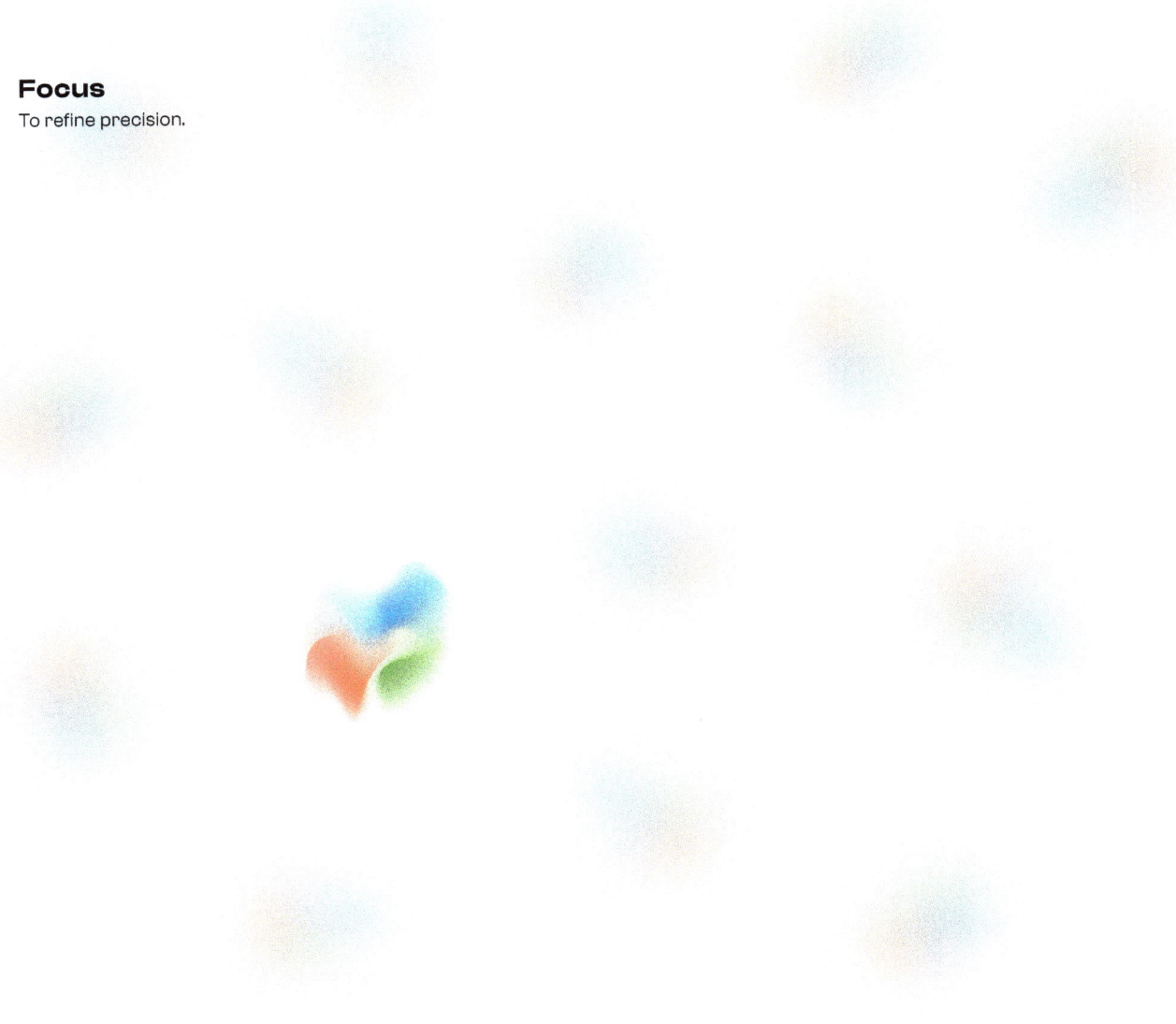

Here's the exciting part: with practice, we can intentionally control and improve how we direct our attention and sharpen our focus. Techniques such as mindfulness and meditation act like workouts for our minds, helping us deliberately fine-tune attention and clarity. These practices allow us to transition from reacting automatically to actively and purposefully engaging with our surroundings.

Consider mindfulness meditation. When you focus on the sensation of your breath, you're zooming in, giving your attention to a single, steady signal. As thoughts arise, such as "What should I make for dinner?" or "Did I send that email?" you gently redirect your focus back to the breath. This practice is not only about redirecting attention; it's about consistently updating your internal beliefs and predictions. As highlighted in recent research, mindfulness meditation can improve your capacity to refine these predictions, making your mental models more accurate and adaptable. Over time, this repeated updating strengthens your ability to maintain clarity and helps you avoid becoming stuck in outdated or unhelpful patterns of thought.

Zoom and focus don't just shape personal perception; they extend outward, influencing collective and cultural narratives. On a societal level, shared attention guides what issues, values, or stories become priorities. When communities collectively sharpen their focus toward common goals, such as investing in critical infrastructure or reducing child poverty, they create a stronger sense of unity, purpose, and coherence.

Conversely, when attention becomes scattered or misaligned (as seen with sensationalized media or fear-driven narratives), it can lead to increased uncertainty, confusion, and anxiety. Intentional attention, practiced individually and within communities, helps foster empathy, mutual understanding, and meaningful progress.

Zoom and focus are key components of the active inference cycle, a process through which we continuously update our beliefs and actions based on new information. Honing the ability to prioritize (zoom) and clarify (focus) signals from both internal and external environments aligns perception more closely with reality, reduces uncertainty, and strengthens adaptive responses to life's changes.

## SPARK

- During your next conversation, notice where your attention naturally drifts. Are you fully engaged, or does your focus wander? What is the energy of the exchange (rhythm, tempo, or tension) and how might you move with or shift it?

- Afterward, reflect briefly on how your attention (zoom) and the clarity of your focus (precision) shaped the interaction. Did noticing those shifts create a different rhythm or nudge the conversation toward a subtle, intended flow?

# THE MARKOV BLANKET: A COMFORTING CONCEPT

# THE MARKOV BLANKET: A COMFORTING CONCEPT

We now understand that we have the opportunity (and the benefit) of deepening our self-knowing. Doing so better equips us to navigate the uncertainty ahead. In the previous chapter, we explored how this unfolds within us, through the interplay of mind, body, and environment. But this same process operates at different scales, using a concept known as the Markov blanket.

At first, the term Markov blanket might sound complex, as though it introduces an entirely foreign concept. In reality, you likely already have an intuitive grasp of the idea—you just might not have named it yet. If you've ever considered how your body is made up of cells, or maybe studied a cell under a microscope in school, you've already thought about the principle behind a Markov blanket.

Imagine shifting your perspective from yourself as a whole person to focus on one of the cells in your body. That cell has its own boundaries and interacts with the world around it (like absorbing nutrients and expelling waste) while maintaining its internal stability. It has a kind of "skin" or layer separating it from the rest of the body, defining what is "inside" the cell and what is "outside." This boundary is its Markov blanket, enabling it to interact with its environment while preserving its own integrity.

Now, imagine zooming out instead of in by thinking of yourself as a person within the larger system of Earth. In this view, your Markov blanket becomes the boundary between you and your environment, helping you interpret, respond to, and regulate your relationship with the world around you. The way we see, feel, and interact with the world means most of us are experiencing and perceiving the world from this perspective most of the time. Consider zooming even further out. As humans, collectively, we are part of the Earth's ecosystems, interconnected through shared resources, climate, and social structures. From this perspective, our species as a whole can be seen as having a kind of Markov blanket, where interactions occur at the boundary between humanity and everything else.

Scientifically speaking, all three perspectives are real and equally valued: my cell in my body, me as an individual, and me as a member of the human species. However, it's my personal experience that makes my perspective center stage.

I find comfort in this concept for several reasons. For example, I have white blood cells, which I'm deeply grateful for. They keep me alive by identifying and attacking harmful pathogens, protecting my body from illness. However, when one of my white blood cells dies, I don't die. I don't even feel pain. My other white blood cells continue their work, and my body as a whole remains in balance. I can be grateful for the existence of that one cell and for its contribution to my health, while also recognizing that its passing doesn't disrupt the greater system of my life. To me, this offers a deeply reassuring way to think about mortality—like stepping back far enough to see my life as one note in a vast, ongoing symphony. Change and death don't silence the music; they help carry it forward. I may be small, but I belong to something enduring and alive.

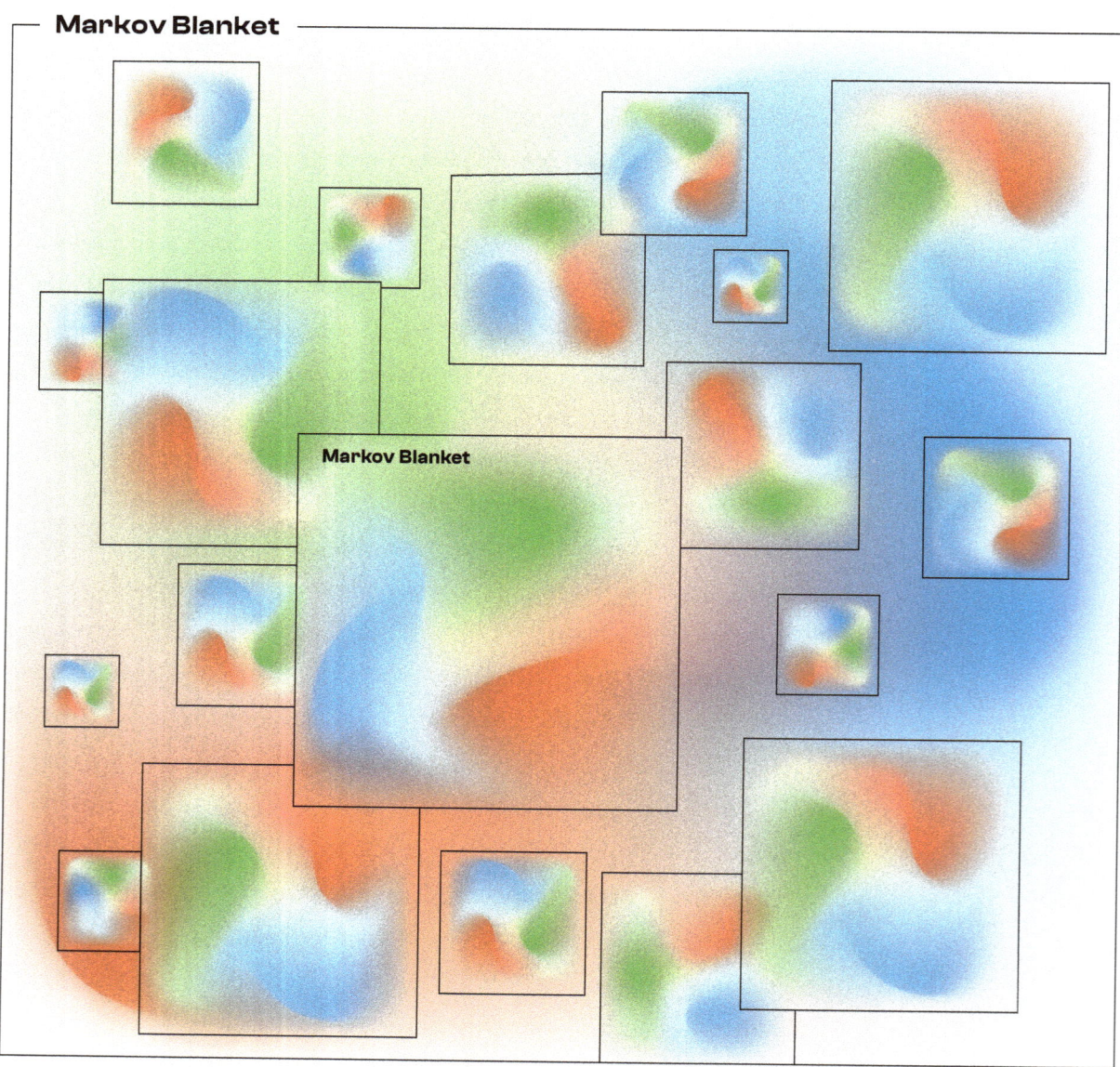

Markov Blanket

Thinking about Markov blankets at different scales (and the idea of Markov blankets within Markov blankets) can illustrate the seamless, interconnected way we self-organize at cellular, personal, and societal levels. For instance, for my cells to thrive, I need to maintain the integrity of my body's Markov blanket, ensuring a stable environment. On a larger scale, for me to have the option to raise children in a safe and supportive environment, society itself must uphold its own Markov blankets and the systems, structures, and cultural norms that provide sustainable and nurturing environments for families. This interplay extends well above and below these immediate experiences to encompass systems like atomic structures and galaxy clusters.

In a similar way, the Markov blanket reminds us that boundaries (whether cellular, personal, or planetary) are not just barriers, but also facilitators. They enable interaction, adaptation, and resilience, allowing systems to persist even amidst constant change. This dual role of boundaries, as both protectors of internal integrity and facilitators of external connections, underscores their dynamic nature and the need for balance between the external and internal.

These ideas play out in daily life. Life unfolds within Markov blankets and societal structures, which are shaped in part through our own actions. There is a kind of circular causality at play. The systems and cultural norms around us shape our behavior and guide what's considered appropriate, possible, or even visible to us (top-down causation). At the same time, our individual actions ripple outward, co-constructing and reinforcing those very structures (bottom-up causation).

For example, we grow up speaking the language we hear around us, but over time, the way we speak and use our slang, expressions, and even typos can gradually reshape the language itself. Similarly, in school, students follow rules set by teachers and administrators, yet how students behave often leads to those rules being revised or new ones being created. These are small, everyday examples of how top-down influence is met with bottom-up adaptation. In this sense, we're not just passive participants in society; we're active contributors helping to maintain and reshape its Markov blanket. The boundary between "me" and "society" is not a wall—it's a living, breathing interface, and we have opportunities to shift that experience for ourselves and others.

These principles, rooted in science, offer a comforting perspective on life's complexities. It highlights how each part contributes to a greater whole while maintaining its identity, demonstrating the balance between individuality and interconnectedness. The Markov blanket serves as a reminder that we are co-creators, both separate and part of something larger—a balance that allows for adaptation, connection, and resilience in the ever-changing dance of life.

## SPARK

 What comes up for you when you reflect on your place within the Markov blankets? When you consider that your body is made up of cells with their own boundaries—each with a role, a lifespan, and a connection to something larger—what do you feel? How does it stir a sense of peace, purpose, or urge to shift something in your life or in the world around you?

"The Markov blanket reminds us that boundaries (whether cellular, personal, or planetary) are not just barriers, but also facilitators.

# UNCERTAINTY AS OPPORTUNITY: BENEFITS OF THE UNKNOWN

# UNCERTAINTY AS OPPORTUNITY: BENEFITS OF THE UNKNOWN

Uncertainty isn't inherently bad. In fact, it's a key opportunity for growth and adaptation. Without it, life becomes boring, monotonous, restless, and disconnected. Instead of viewing uncertainty solely as something to avoid, we can reframe it as an opportunity to gather useful information and foster balance. This shift in perspective allows us to embrace uncertainty as an opportunity to refine our understanding of the world and ourselves, transforming it from a threat into a pathway for growth and vitality.

Opportunities to resolve uncertainties can be divided into two broad categories: utilization and exploration. Utilization involves using existing knowledge to navigate familiar situations effectively. For example, consider cooking a meal you've made many times before. You know the recipe, the steps, and the expected outcomes, making the process predictable and stable. Each time you cook, you reinforce your ability to handle this specific situation, building confidence and efficiency over time.

Exploration, on the other hand, involves seeking out new experiences or information to expand your understanding of the world. Imagine trying a completely new recipe at a friend's house. The cookware might be unfamiliar, the ingredients new, and the added dynamic of social interaction might feel

**ANNOTATION**

The scientific term often used in this context is exploitation, without its usual negative connotation. However, we've chosen to use the word utilization for flow and relatability.

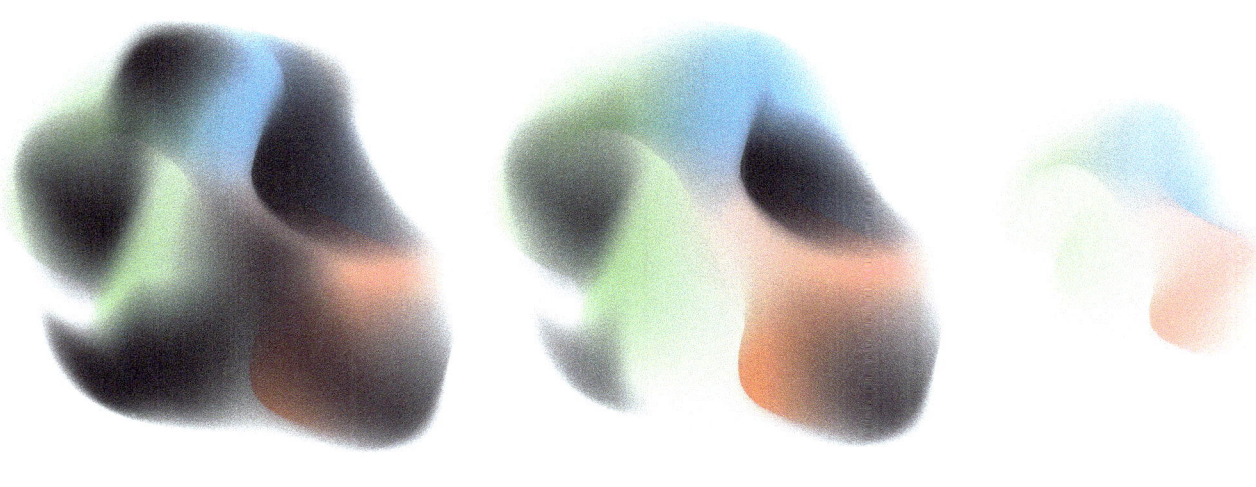

Active inference cycle with
**overwhelming levels of uncertainty**

Active inference cycle with
**balanced uncertainty**

**Bored** active inference cycle

uncomfortable at first. However, it's through this initial uncertainty that growth occurs. Experimenting with new actions gathers fresh insights and improves your ability to adapt to novel situations, enhancing both skill and adaptability across various areas of life.

Balancing familiarity with novelty is key for responding effectively to uncertainty; without this balance, we risk stagnation in outdated cycles or becoming overwhelmed by constant change. Intentionally incorporating new activities into one's life fosters this balance, helping to build comfort with discomfort. This process is supported by key mechanisms in the brain. For example, repeated exposure to uncertainty or feared situations activates the prefrontal cortex, which in turn regulates the amygdala. This process aids fear extinction and helps new experiences feel less threatening over time. Additionally, habituation (the brain's ability to normalize new experiences) reduces the discomfort of novelty, fostering resilience and adaptability.

Engaging with uncertainty can also activate the brain's reward system, releasing dopamine and reinforcing the positive effects of exploration and adaptation. Over time, these neural processes enhance emotional stability, improve specific skills, and equip the brain to respond more effectively to unexpected challenges. This strengthens our resilience and builds confidence in moving through an inherently uncertain world.

However, building these habits and inciting shifts requires both time and effort, as they involve physical changes in the brain. Taking small, manageable steps to incorporate new habits supports these physiological processes. For instance, setting aside time for novel activities in a structured and intentional way creates opportunities for meaningful growth without overwhelming the system. Removing the internal pressure for initial mastery allows room for growth and learning. Research consistently shows that successful habit formation begins with incremental steps. Developing comfort with the discomfort of doing something new doesn't require mastery on day one; rather, it involves identifying and taking achievable steps that align with the natural rhythms and capabilities of our physiology.

Our brains are constantly making predictions about the world around us based on prior experiences and beliefs. Intentional exploration and practice expand this library of knowledge, laying a foundation for greater coherence and stability throughout our daily lives. Without these updates, we risk getting stuck in cycles of outdated information that fail to resolve the uncertainties we encounter. Cultivating a lifestyle that incorporates new experiences ensures we remain adaptive, effectively balancing utilization and exploration while equipping ourselves to navigate life's challenges and opportunities with greater clarity and confidence.

## SPARK

- Consider how you might invite more newness into your life in a way that supports your overall system.

- What kind of activity would serve you best right now? Maybe it's something lively and stimulating, like visiting a museum or joining a community event. Or perhaps something slower and calming, like floating in a sensory deprivation tank or walking through a quiet part of town you've never explored.

- What's a realistic implementation for you? Do you need to schedule it (carving out time intentionally on your calendar) or does it feel better to stay open and follow spontaneous nudges?

- Whatever you choose, the intention is simple: to foster doing something unfamiliar and observe how it feels.

# AN EMPATHETIC LENS: HOW SURPRISE & UNCERTAINTY AFFECT CONNECTION

# AN EMPATHETIC LENS: HOW SURPRISE & UNCERTAINTY AFFECT CONNECTION

Human connection is essential to reducing uncertainty in our lives. Whether through family, friendships, or community, these relationships provide consistent and predictable sources of emotional regulation. A key component of this process is empathy. Often thought to be supported by mirror neurons, empathy allows us to simulate and predict the actions and emotions of others. A favorite discovery story of mine, mirror neurons were first discovered in the 1990s by researchers studying monkeys. It was first inadvertently observed when a monkey's brain unexpectedly responded to an action performed by a researcher. The monkey was already hooked up to monitor its brain activity for a set of planned tests. However, when the researcher picked up an object and the monkey remained still (a condition not part of the initially planned testing sequence), the same neural pattern fired as if the monkey itself had performed the action. This led to the inadvertent discovery of mirror neurons. These specialized brain cells activate both when an individual performs an action and when they observe someone else performing the same action. In other words, parts of our brains react the same whether we're experiencing something directly ourselves or empathizing with someone else.

Building on the innate capacities of our mirror neurons, this section weaves

**ANNOTATION**

The story of how mirror neurons were discovered is especially inspiring to me because of their accidental nature. It makes me wonder: how many phenomena remain unknown to us, not due to a lack of technology, but simply because we haven't yet thought to look?

in elements of active inference and uncertainty. Taking a moment to reflect on how the free energy principle shapes human behavior can deepen our empathy, helping us navigate interactions with greater effectiveness. This supports those around us while strengthening the predictability and consistency of our own connections.

## Discomfort of the Unexpected

One key concept often helpful to understand is that unpredicted discomfort is often more distressing than expected discomfort because the unexpected in itself is often uncomfortable. For example, picture a calm drive suddenly interrupted by a child in the backseat unexpectedly hitting another child. The child who is hit experiences not just physical pain but an added layer of distress from the unexpected nature of the action. This surprise amplifies their discomfort, causing them to react by hitting back with greater force. Rather than a deliberate act, the escalation arises instinctively in response to the discomfort from the unexpected hit.

This principle extends to our social interactions. Consider a family member who makes a hurtful comment during a conversation. While the words sting, the underlying cause might be their own struggle with uncertainty, perhaps about their role in your life or their sense of stability in the relationship. For instance, an often otherwise supportive parent might discourage you from pursuing a risky career path, not out of malice, but because their uncertainty about your future success triggers their need to protect you and themselves. The unexpected lack of support can make the interaction feel even more painful. Understanding this dynamic through empathically considering their motivations can help us hold space for their experience while also attending to our own needs, rather than reacting in ways that further erode the connection and ultimately our own sense of stability.

## Sensory Attenuation

There's a second concept that often contributes to escalation: the brain filtering out sensory inputs during action, a process known as sensory attenuation. To act effectively, we often temporarily ignore the immediate sensory feedback of our actions. For instance, when preparing to stand up from a chair, the brain must reduce the sensory evidence that we are still seated; if the brain didn't temporarily set aside the sensation of sitting, it would struggle to initiate the action of standing because it'd be contradicted by the sensation of sitting. Sensory attenuation helps prioritize certain signals or intentions over others, enabling smooth initiation and execution of actions. In extreme cases, we can see how this process can break down, like Parkinson's disease, where the brain's inability to suppress sensory feedback can contribute to difficulty initiating movement.

The selective focus of sensory attenuation creates an asymmetry in perception during interactions. For instance, let's revisit the scenario of the two children in the backseat of the car. When one child hits another in the backseat of a car, sensory attenuation causes the first child to perceive their action as less intense than it actually is. When the other child hits back, however, they are fully

Reflecting on my most regrettable family and work experiences, I've noticed that surprise often played a pivotal role. Whether initiated by myself or others, the unexpected frequently triggered a chain of escalating responses. As a result, I now approach situations involving unexpected elements with greater caution—working to minimize unnecessary surprises for others and prioritizing de-escalation whenever possible.

aware of the impact, leading them to feel as though the response was dispro-portionately stronger. This false perception can also escalate the conflict as each person believes they are simply responding in kind. If unchecked, this cycle esca-lates while eroding the possibility of reciprocal understanding.

This pattern of erosion and escalation can take many forms. One common example is a conversation that intensifies due to misunderstandings or perceived overreactions. At its extreme, this dynamic leads to dehumanization, where the other person is no longer seen as fully human because their behavior seems fun-damentally different from our own. This can prevent constructive dialogue and deepen divisions. When this happens, it poses significant risks by escalating con-flicts, fostering alienation, and breaking down social cohesion. Incorporating awareness of this dynamic into active inference cycles provides opportunities to reduce escalation and foster constructive dialogue and mutual understanding.

## The Narrowing Effect of Threat and Uncertainty

When we feel threatened (whether physically, socially, or psychologically) our minds respond by narrowing our attention and clinging to familiar beliefs and actions. Recent scientific research describes this as attentional narrowing, a natural brain response driven by stress and uncertainty. Attentional narrowing helps explain why, under pressure, we often default to habitual patterns, even if they aren't the best choices available to us.

The neuroscience behind this response is revealing. When uncertainty spikes, our brains release chemicals like adrenaline and noradrenaline.

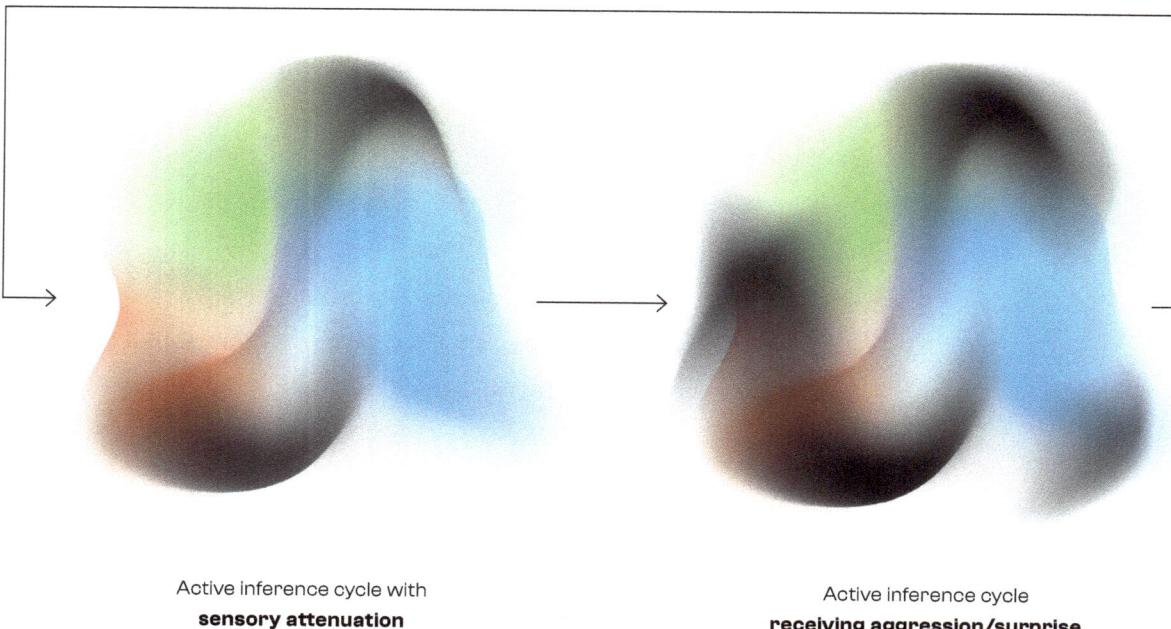

Active inference cycle with
**sensory attenuation**
**being an aggressor**

Active inference cycle
**receiving aggression/surprise**
**(no sensory attenuation)**

These neurochemicals disrupt the normal functioning of the prefrontal cortex, the region associated with thoughtful decision-making and flexibility. Simultaneously, "neural gain" (the amplification of strong signals and suppression of weaker ones) causes familiar ideas and actions to dominate, effectively sidelining alternative perspectives.

This process has crucial implications for understanding political polarization. Consider political disagreements, which are often fueled by perceived threats such as economic instability, social uncertainty, or identity conflicts. The heightened uncertainty involved in these situations narrows our focus dramatically, making us resistant to alternative viewpoints and overly committed to

Active inference cycle
**delivering equal
aggression/surprise**

Active inference cycle
**receiving aggression/surprise
(no sensory attenuation)**

previously held convictions. Such rigidity manifests in intensified ideological positions and reduced openness to dialogue, fueling cycles of conflict rather than resolution.

One telling example is how uncertainty in one domain of our lives, such as personal economic anxiety, can spill over and reinforce rigidity in unrelated areas, such as ideological beliefs or political views. Research demonstrates that relationship uncertainty or financial instability can intensify our ideological convictions. When faced with uncertainty, our brain seeks comfort in certainty—even if that certainty involves polarized or extreme beliefs. This insight helps explain why, in times of societal upheaval or perceived threats, communities often see rising ideological divides, as individuals reflexively gravitate toward the reassuring familiarity of rigid viewpoints.

Understanding this automatic cognitive mechanism provides us with a powerful tool for empathy and communication. When we see rigidity in ourselves or others, particularly in emotionally charged contexts like politics, it's beneficial to recognize it as a natural reaction to uncertainty rather than stubbornness or ill intent. Identifying and naming this reaction makes it easier to manage emotional responses, staying open and empathetic instead of reactive and defensive. It enables us to consciously choose dialogue and flexibility over entrenched division, creating pathways toward understanding and connection even amidst uncertainty.

Another method I frequently employ, and which is woven throughout this book, is reminding myself and others that despite differing political views, religious beliefs, or cultural backgrounds, we typically share a large percentage of the same fundamental beliefs and values. Even with someone whose religious or political perspectives starkly contrast with our own, we likely agree on foundational principles: that murder is wrong, that children deserve nurturing and protection, that driving on a consistent side of the road prevents accidents, and that language is essential for effective communication. Thus, when presented with the opportunity to discuss how we might successfully navigate the uncertainty before us as a society, we can approach these conversations in a way that acknowledges existing differences but emphasizes shared perspectives.

From a scientific standpoint, this intervention is effectively shifting our

"Unpredicted discomfort is often more distressing than expected discomfort because the unexpected in itself is often uncomfortable.

cognitive processing from a narrow, threat-focused perspective to a broader, integrative viewpoint. A constricting approach might be to begin by saying, *"Historically, humans haven't been able to handle so much change in one lifetime; we just need to dodge bullets and wait for things to improve."* These types of responses, as we've seen from the science above, often lead to the brain narrowing its attention, fixating on differences and disagreements.

In contrast, a more expansive way to initiate dialogue could sound like: "We all experience varying levels of fear given the significant uncertainty we face today. Yet we also recognize that combining our diverse perspectives provides us with a richer, more complete picture. Considering your own beliefs, what do you feel is the most helpful and pressing action we can take to navigate this next transition effectively? How can we thoughtfully integrate our differing views to chart a course that, while perhaps not perfectly aligned with every individual's ideals, helps us avoid catastrophic pitfalls we collectively wish to prevent?" Intentionally zooming out to consider shared beliefs counteracts automatic narrowing, re-engages higher-order reasoning, and encourages neural flexibility. This cognitive shift decreases stress-induced rigidity, promoting enhanced creativity, improved problem-solving capabilities, and greater openness to collaborative interactions. Essentially, recognizing our shared humanity and common values mitigates uncertainty-driven defensiveness, creating psychological space for empathy, innovation, and constructive dialogue.

Ultimately, recognizing the cognitive and emotional processes underpinning attentional narrowing helps us foster compassion for ourselves and others when navigating complex and emotionally charged topics. This awareness can break cycles of escalation, allowing space for more constructive and empathetic exchanges, especially in areas such as political discourse, where uncertainty and perceived threats are often high.

## Tying Back in Empathy

Now that we understand how the discomfort of unexpected surprises can amplify distress and how the selective focus of the brain during action can inadvertently escalate conflict, we can return to the foundation of empathy. The mirror neurons we described earlier provide us with the wiring to use empathy for our own benefit. They contribute to the opportunity for a crucial balance, acknowledging another's perspective while honoring our own needs. It allows us to navigate interactions with curiosity and understanding, shifting our focus from judgment to connection and fostering deeper relationships. It shifts our focus from judgment to curiosity, helping us see actions not as isolated behaviors but as attempts to reduce uncertainty. **Empathy doesn't excuse harmful actions but creates room to understand the motivations behind them, fostering deeper connection and understanding.**

At a societal level, this concept helps us navigate differences in how people optimize their lives. Modern life often pushes us to optimize for societal markers such as wealth, success, and status, but these don't always align with what our systems truly need. The free energy principle reminds us that our foundational drive is not external validation but coherence and reduced uncertainty. We can also combine an empathic lens with our understanding of how people's desire to reduce uncertainty is shaped by their past experiences and personal circumstances. For instance, someone who grew up in financial instability might prioritize saving over creative pursuits because their system equates money with safety. Another person, raised in an emotionally chaotic environment, might seek predictable routines to feel secure. These choices reflect not flaws but adaptations to their circumstances and internal needs. Recognizing this shifts us from judgment to understanding, opening space to empathize with others' attempts to navigate their worlds.

When we approach others through this empathetic lens, we acknowledge that every behavior is an effort to reduce uncertainty, shaped by unique experiences and environments. We also start to see how our inherent programming, such as the heightened discomfort from unexpected hurtful behaviors or the brain's tendency to filter sensory inputs during actions, can distort our understanding of situations. This awareness helps us build bridges across differences, fostering deeper compassion and connection while also balancing our own needs. Ultimately, this creates greater stability for ourselves in our environments, strengthening our relationships and stabilizing our shared humanity.

**ANNOTATION**

I often find this especially helpful when considering dynamics at the geopolitical level.

## SPARK

- Think of a moment when you found yourself caught up in a challenging interaction or situation that felt like it might have escalated in a distressing way, perhaps a disagreement with someone close to you or a tense encounter at work. At the time, you likely acted in a way that felt protective or justified.

- Looking back with your deeper understanding of how cycles of tension and free energy work, consider reflecting on opportunities to have responded differently. How could you have balanced protecting yourself and your needs, while also being intentional about not escalating the tension and free energy further?

# INTENTIONS & THE BIDIRECTIONAL NATURE OF THE BRAIN

# INTENTIONS & THE BIDIRECTIONAL NATURE OF THE BRAIN

If empathy helps us connect with others, intention helps us reconnect with ourselves. But before diving into specifics, it's useful to address how the brain interacts with the world. Contrary to the dated notion that our brain passively responds to the environment (a one-way transfer of information), modern neuroscience reveals a bidirectional process: the brain not only receives information from the environment but also actively projects expectations outward. For instance, when I reach for my water bottle on my desk, my brain anticipates its location and weight before I even touch it. If the bottle isn't where I expect or it feels unexpectedly empty, this mismatch between my expectation (outward arrow) and reality (inward arrow) forces my brain to update its interpretation and adjust my actions accordingly. We don't notice this effect very often because we're so good at seamlessly predicting and updating, but if you were to sneak up behind me while I'm deep in thought writing and move my bottle of water, I'd likely knock it over as my hand reaches for where it used to be, or pause, confused, and have to look up from my work to locate it.

This two-way interaction between expectations and reality fundamentally shapes the power of intentions. We can compare our intentions to mental blueprints to illustrate. They operate within the bidirectional framework. If we

**DEFINITION**

Functioning in two directions

**Navigating Uncertainty**

remember the cycle that is active inference, then one arrow at all times is pointing outward (what we expect) and another is pointing inwards (what we perceive and experience), helping to bring coherence and reduce the surprise as we navigate the world.

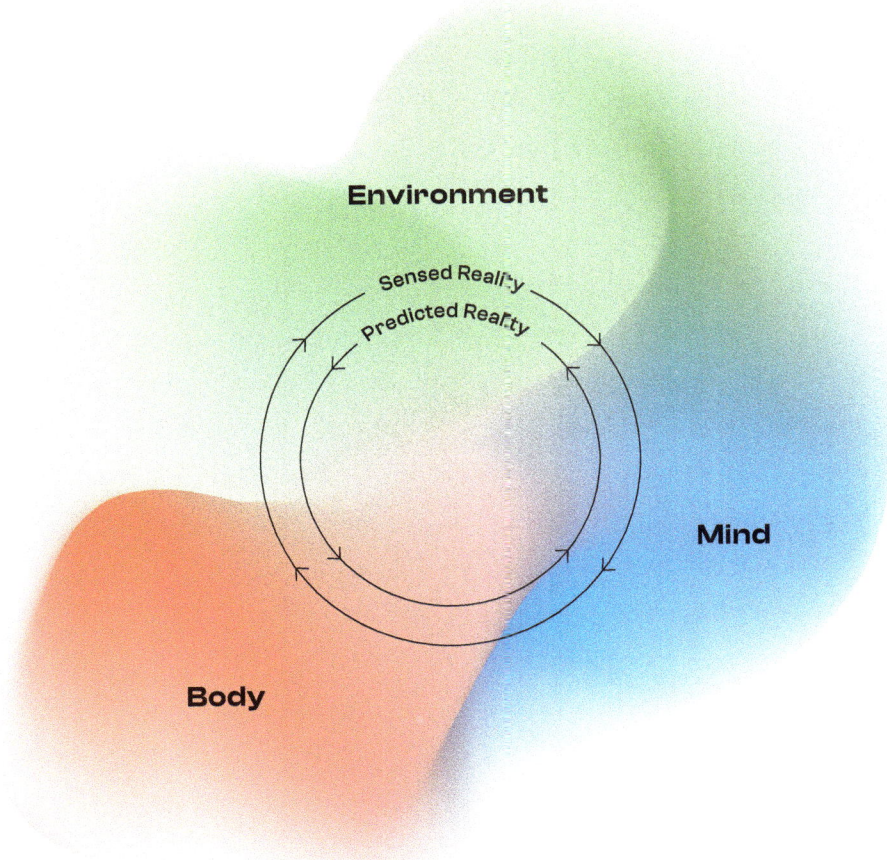

The key point here is that the influence of the outward arrow of expectations is not just inward from what we perceive and experience. We actively shape the world as we experience it. Our expectations aren't just predictions; they are active forces that shape what we see, hear, and experience. In a very real way, we author our perceptions by acting on the world to bring our expectations to life. And in turn, the world pushes back, refining and reshaping those expectations. It's a dynamic dance: we make reality conform to our expectations, and our expectations adapt to reality. Thus, the intentions we set affect our overall experience in a deep and automatic way.

Our mental blueprints are, in part, formed by our beliefs about who we are and what we aim to achieve. Scientists often describe intentions as "priors over policies," meaning they are pre-existing beliefs that guide possible actions. These existing beliefs are shaped by our identity and anticipated outcomes, which explains why intentions often feel deeply personal and morph over time. For example, if you see yourself as a creative individual, your intentions will likely revolve around actions that align with that identity, such as setting aside time to paint, write, or engage in artistic hobbies. This self-model—the narrative of who you are—acts as the foundation for your intentions, influencing not just what you do but how you perceive the outcomes of your actions.

Because of these bidirectional arrows, our intentions also play a critical role in shaping our active inference cycles and navigating the uncertainty within them. Let's consider how intentions stack up against what we know about reducing dissonance, the free energy, in our overall system. Intentions aligned with reducing system dissonance (rather than shallower societal ones) can better serve our systems at a fundamental level by resolving internal and external tension more effectively.

Think about your intentions across the domains of mind, body, and environment. How much do they reflect what we know about our deeper needs, as opposed to what society advertises will bring happiness? Is there an opportunity to shift your intentions to align more closely with your intrinsic system needs? Intentions are deeply personal, so I can't tell you what yours should be. However, I share mine here as an example and perhaps as inspiration for a new way to look at yours. For instance, I used to work out as a way to appear fit so others would perceive and value me as a fit person. However, this intention carried high levels of free energy because I couldn't control what others thought of me, and it didn't address my deeper need for stability and coherence. Since understanding free energy and active inference, I now work out to increase my body's capacity to support my active inference cycle. This intention carries inherently less free energy-dissonance because it's grounded in deeper science and my fundamental needs, rather than external factors beyond my control. It also creates a positive feedback loop within my active inference cycle.

If we delve deeper into this example, we can see it reflects the larger dual imperatives of exploration and utilization we've touched on throughout the book. These imperatives align with two types of intentions: pragmatic intentions (based on what is already known) and epistemic intentions (focused on seeking new knowledge). Pragmatic intentions aim to maximize expected outcomes using established information. For instance, I know my body responds well to a mix of yoga, barre, and light running. Too much heavy lifting, and I tend to sustain an injury. So, I set pragmatic intentions to maintain a realistic balance of these activities, adjusting based on the specific needs of my overall system at any given time.

At the same time, I recognize that incorporating occasional weightlifting has been shown to be beneficial for long-term age-related mobility. Here, I set an epistemic intention: to explore the gym near my house and experiment with incorporating occasional weightlifting. I don't yet know what this routine will look like or how I'll need to structure it to avoid injury, but I've set the intention to start exploring this over the next month or two. Epistemic intentions like this allow me to reduce uncertainty by refining my understanding. In this case, it's about what type of weightlifting regimen could fit into my system's overall needs.

Balancing these dual imperatives ensures that my intentions are both practical and open to growth. Pragmatic intentions provide stability by leveraging what I already know works, while epistemic intentions invite curiosity and adaptation. Together, they help align my actions with my deeper needs, allowing me to navigate life's challenges with greater resilience.

When we shift our focus from the body to the mind, the dual direction of information flow becomes just as relevant. This understanding has significantly influenced my intentions for my mind. For instance, it has reshaped the way I engage with media. Rather than consuming content purely for entertainment, I now carefully consider how the material I engage with shapes my mental

" **Our expectations aren't just predictions; they are active forces that shape what we see, hear, and experience.**

expectations of the world. I prioritize content that reinforces accurate and mindful outward projections (reducing free energy), which has led to changes in how I use social media, the types of movies I watch, how I interact with AI, and the books I choose to read. These shifts reflect a balance between building on familiar processes and exploring new, less certain ones, fostering a dynamic cycle of growth and alignment.

Finally, I consider my environmental intentions as a balance between familiar, productive, and restorative spaces, and novel, information-rich, and challenging ones. This thoughtful balance nurtures the dynamic alignment and growth of my mind-body-environment cycle, fostering resilience and adaptability in an ever-changing world. Beyond personal growth, I have a deep desire to help others, whether it's supporting my cousin as she welcomes a new life into the world or creating this book with a genuine intention to provide a valuable resource for others. My intention is to strengthen the interplay of supporting others, recognizing that by investing in my own well-being, I cultivate a more capable and resilient system to extend that support outward.

By the time you're reading this book, these intentions have likely morphed. Intentions are not static; they are part of a dynamic feedback loop. As we act on our intentions, we refine them based on the outcomes we experience. This iterative cycle ensures that our intentions remain flexible and responsive to changing circumstances, supporting growth and resilience.

## SPARK

✦ Consider your domains of mind, body, and environment. What is your key intention within each? Are there ways you might revise or shift your intentions to reduce dissonance or free energy—by aligning more closely with the limits of your control over your surroundings and other people, the physiology of your body, or the bidirectional nature of the brain?

# WILLPOWER VS. VOLITION: A MORE EFFECTIVE USE OF OUR FINITE ENERGY

# WILLPOWER VS. VOLITION: A MORE EFFECTIVE USE OF OUR FINITE ENERGY

As we deepen our understanding of active inference, we can explore how we turn mental blueprints into embodied action—not by sheer force, but through a more sustainable and adaptive mechanism than willpower alone. As we go about life, our energy is finite, and our ability to mobilize it through willpower has limitations. This isn't because we're not trying hard enough, but because **intrinsic constraints within our mind-body-environment shape what's possible.** Reflecting on the concept of willpower reveals a clearer view of its limitations and explores a shift towards volition, more effectively utilizing our finite resources.

Let's consider the whole system: mind, body, and environment. Willpower, as it's traditionally understood, resides in the brain. This means that by relying solely on willpower, we're drawing on just one-third of our overall potential. Then, within the brain itself, things get even more specific. The brain is incredibly complex, with distinct regions responsible for different functions. Notably, only about one-third of our brain mass (the prefrontal cortex) underpins willpower. So now, we're down to a third of a third. And that's not the end of it. The prefrontal cortex is performing multiple key tasks at any given moment, from problem-solving and decision-making to regulating emotions and managing social interactions. Willpower is just one of many functions it's juggling. Illustratively speaking, we're

**Navigating Uncertainty**

operating at a fraction of a fraction of our total system's capacity. By relying exclusively on willpower, we limit ourselves to an incredibly small portion of what our mind-body-environment system is capable of achieving.

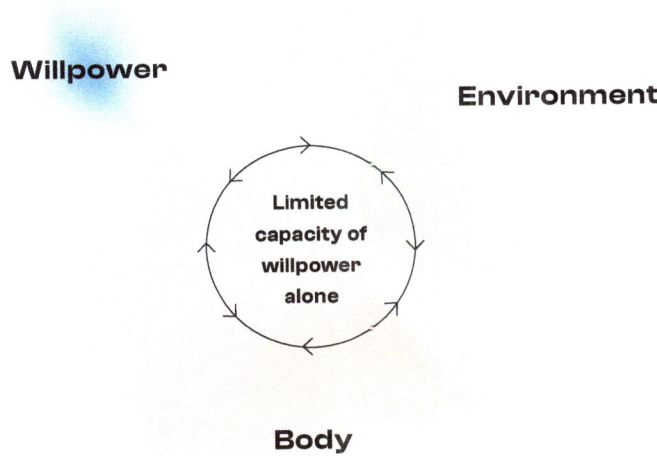

Beyond this, other systems in the brain can either hinder or support that limited willpower. For example, the limbic system, which is responsible for our immediate responses and emotional reactions, operates largely outside of conscious control. In other words, we largely can't control those immediate emotional reactions we feel. When we're tired, stressed, or overwhelmed, the reactive systems dominate, reducing the influence of the prefrontal cortex. Thus, when willpower fails, it's not because we're weak or lack discipline; it's because our systems are working exactly as wired.

Instead of over-relying on willpower, we can shift to volition—the ability to align our actions with our beliefs and goals. This helps us leverage our entire system to achieve greater coherence and flow. However, volition is not about brute force; it's about working smarter, not harder, by aligning all parts of our system to act in concert. Let's explore this difference more deeply.

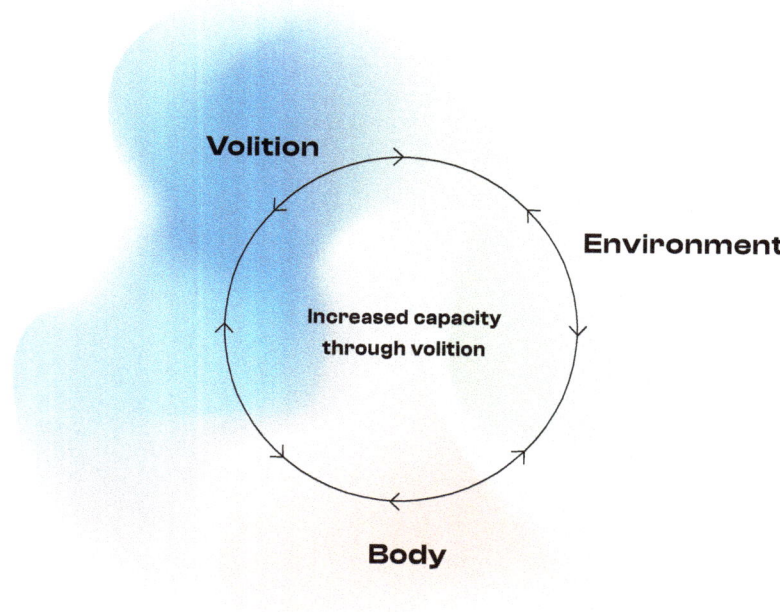

## Understanding the Role of Volition

Imagine someone who's exhausted after a long week at work. If they rely solely on willpower, they might label themselves as lazy and push through their exhaustion, further depleting their energy and leaving little capacity for their family or personal well-being. In contrast, volition offers a more adaptive approach. When fatigue is seen as a natural response and beliefs adjust accordingly, space opens to reassess and reorganize tasks. Additionally, their system experiences the additional relief of not being burdened by the inaccurate perception (free energy) of their supposed laziness. This allows them to balance critical demands with the body's need for rest, aligning actions with their system's deeper needs and fostering a deeper sense of coherence and sustainability.

## The Physiology of Misalignment

Physiologically, chronic misalignment between our intentions, actions, and environment can also have significant consequences on willpower. Firstly, it triggers the body's stress response, activating the sympathetic nervous system and elevating cortisol and adrenaline. Chronic activation can lead to fatigue, cognitive impairments, and compromised immune function. Secondly, persistent misalignments result in prediction errors within our brain, demanding considerable physiological energy to resolve these discrepancies, leading to exhaustion. Thirdly, ongoing stress responses disrupt key brain regions, impairing emotional regulation and cognitive clarity, which perpetuates cycles of ineffective behavior. Finally, misalignments disturb our body's essential homeostasis, forcing the system to repeatedly expend energy attempting to regain balance. Thus, alignment is physiologically essential, helping to reduce internal tension while boosting overall resilience and health.

## Self-Efficacy: A Tool for Volition

Self-efficacy refers to the belief in one's ability to execute actions and influence outcomes. Originally proposed by psychologist Albert Bandura, this concept is supported by decades of research across psychology, education, and health. People with high self-efficacy tend to persist through challenges, adapt more effectively to change, and recover more quickly from setbacks.

In many ways, self-efficacy serves as a volitional tool. It's the internal signal that we are capable of acting in line with our goals, especially under pressure. While willpower attempts to override difficulty through short-term force, self-efficacy helps sustain action by reinforcing our belief that change is possible and meaningful. This makes it a critical bridge between intention and execution. Rather than relying solely on a narrow slice of the brain (as willpower does), self-efficacy draws upon embodied experiences, environmental cues, and internal narratives to generate momentum from a deeper, more integrated place.

From an active inference perspective, self-efficacy reflects the system's confidence in its ability to resolve prediction errors and reduce free energy through intentional action. When someone believes they can act effectively, their system is more likely to engage in exploration and adaptive updating, leading to greater coherence across the mind, body, and environment. For example, a parent facing a child's behavioral challenge may feel unsure at first, but a strong belief in their own capacity to learn new parenting tools can drive them to seek guidance, try new strategies, and remain emotionally present through the process. This belief helps align their internal expectations with external realities, making adaptation and coherence more likely across their mind-body-environment system. In this way, self-efficacy functions as both a predictor and product of successful volitional alignment, supporting long-term adaptability rather than reactive control.

Personally, rekindling my self-efficacy was essential in the creation of this research and book. During the development process, external validation was sparse, and uncertainty was high. A key resource that kept me going was the belief that I could adapt in a constantly-evolving ground, and that, even as a limited human, my training in the scientific method could allow me to translate complex ideas into helpful, accessible insights. That internal sense of efficacy became a volitional anchor: not a forceful, unreliable push, but a grounded trust in my capacity to navigate uncertainty with coherence and contribution.

## Balancing Pragmatic and Epistemic Intentions

We've touched on the concepts of pragmatic (what is already known) and epistemic (forging new information) several times already. These ideas can also be applied to volition. Pragmatic volition aims to maximize what we know works, such as relying on a familiar stress-management strategy like taking a walk or practicing deep breathing exercises to address immediate challenges effectively. Epistemic volition, on the other hand, seeks to explore new possibilities. For instance, you might try progressive muscle relaxation or therapeutic creativity to see if these approaches add value.

A common pattern many of us fall into is the overuse of pragmatic willpower, sticking to behaviors we know but that may have diminishing effectiveness, without expanding into epistemic volitions. **Incorporating new activities enhances our capacity to meet life's complexities while making the most of our finite energy.**

## Leveraging the Entire System

Volition takes into account not just the brain but the entire mind-body-environment cycle. While the prefrontal cortex provides direction, the body supports action, and the environment shapes context. By aligning these elements, volition creates a feedback loop where each part reinforces the other. For example, setting up an environment that reduces distractions or fosters positive habits can make it easier to stay on track, lessening the strain on the brain's willpower reserves.

This approach also integrates the limbic system rather than working against it. For instance, instead of resisting emotional impulses, volition acknowledges their presence and reframes them to align with your intentions. Feeling stressed about an upcoming deadline? Instead of forcing yourself to work harder out of sheer willpower, you might set an intention to use that stress as a signal to take a break, pause to consider novel strategies, or seek support, ultimately reducing tension and fostering clarity.

## Moving Toward Alignment

The key to volition is alignment by bringing intentions, actions, and environment into greater coherence. When this happens, your energy is used more efficiently, and your actions feel purposeful rather than forced. Volition isn't about eliminating challenges but about engaging with them in a way that integrates your whole system, creating greater opportunities for growth and fulfillment.

In this way, volition becomes a bridge between intention and action, making the most of your finite energy while reducing the strain on willpower alone. When volition takes the lead, we meet life's challenges with greater ease and draw on the fuller potential of our mind-body–environment system.

## SPARK

One way the tension between willpower and volition often shows up is through the inner critic. Think of a situation when you often notice a recurring voice of self-judgment. What is the underlying challenge your inner critic is attempting to address or solve? Instead of using willpower to silence or override it, explore potential shifts such as adjusting your mindset through self-compassion, supporting your body by honoring its natural limits, or modifying your environment to reduce friction and add supportive elements. These shifts might allow you to respond with greater clarity, alignment, and resilience.

# EMBODIMENT: A COHESIVE EXPERIENCE

# EMBODIMENT: A COHESIVE EXPERIENCE

At the time of writing this book, the word **embodiment** has been gaining recognition, almost reaching rock star status. But what does it truly mean within the scope of our conversation? Here, embodiment means the body takes an active role in shaping how we think, feel, perceive, and act, rather than merely housing the mind. Our bodily states shape our attention, influence our emotions, and help determine how we interpret the world. When we're embodied, we're not just thinking about our experience; we're living through it, with participation of the nervous system, physiology, and sensory feedback. To me, it feels like I am within myself, rather than removed from it. However, this is another area where words often fall short. By both definition and current research, embodiment isn't something we can achieve through cognition alone. It requires the active participation of the body.

A simple way to think of embodiment is being fully immersed in our active inference cycles. For example, imagine noticing subtle tension in your chest during a meeting, as an interoceptive signal might help you address unspoken concerns or navigate the discussion more effectively. At other times, we might distance ourselves from these signals to maintain overall balance, such as choosing not to dwell on a tight chest sensation when it feels overwhelming. Grounding

**ANNOTATION**

While I'm cautious about broad claims, I've only seen embodiment develop through experiential or body-based practice (rather than cognition alone), and I've yet to meet anyone who regretted investing in it through validated practices.

practices can help us reconnect with our cycles, allowing us to process these signals constructively and experience the holistic benefits of being at home within our system. This state of embodiment transforms the body into a **sensory organ**—an essential source of evidence that complements external inputs like vision. These bodily cues enrich our sense-making, grounding our perspective in the body and opening space for pleasant visceral states.

This process is known as **interoceptive inference**, using bodily sensations with cognitive signals to maintain balance and a coherent sense of self. For example, consider a person hesitating over a job offer. Rather than dismissing their hesitation as mere fear of change, they might tune into their body's signals (like a knot in their stomach), which could reveal that the role doesn't align with their deeper intentions. This deeper awareness enables them to make a decision that feels right both mentally and physically. When bodily cues are integrated with mental insights, a more cohesive understanding of the self emerges, actions align more closely with intentions, and overall well-being is enhanced. Without this integration, we risk acting on fragmented perceptions, often missing the mark. Fortunately, perfection isn't necessary. The curve of optimization is forgiving, and even a moderate effort to integrate body and mind yields significant benefits.

Our body is constantly generating interoceptive signals (heart rate, hunger, muscle tension), which the brain interprets to maintain self-knowing. These signals are not peripheral; they are central to how we experience being "ourselves." For instance, I've noticed times when my body signals social anxiety. By approaching these signals with curiosity, I've been able to resolve them more effectively. Sometimes, the discomfort pointed to a real dynamic in the room that required action. Other times, it reflected past experiences that colored my present. By integrating these insights, I've been able to engage with others from a more grounded place, cultivating deeper, richer, and more fulfilling connections.

As we saw previously, emotions provide another lens for understanding embodiment. We're not born with an innate manual for interpreting emotions. Instead, emotions arise from the brain's best guesses about the causes of bodily signals, shaped by context and prediction. Increasing the cohesiveness of our body within the active inference cycle sharpens these inputs, making our emotions clearer and more actionable. Imagine your body as the sensors on a self-driving car. If we dissociate from our bodily sensations, it's like covering those sensors, limiting our ability to navigate the world effectively.

Embodiment also helps us label our states more accurately. For example, feeling depleted at the end of a week might be misinterpreted as laziness without the body's input. Recognizing physical fatigue, however, aligns us with the reality of carrying ourselves through intense demands, reducing unnecessary internal conflict and fragmentation.

Over time, we can refine our ability to interpret our bodily signals. Chronic tension

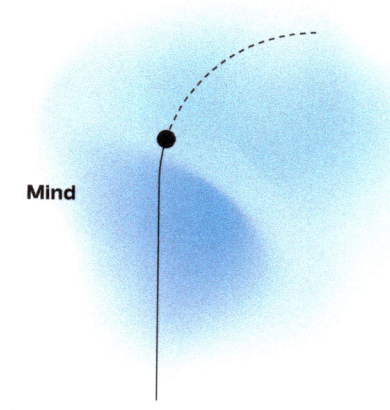

**Cognitive solutions**
with limited optimization
of outcome potential

**Mind**

**Navigating Uncertainty**

patterns, for example, often form outside of conscious awareness. Reintegrating these signals into the active inference cycle reveals their connection to states such as stress, turning discomfort into useful sensory information, and empowering us to respond more effectively. Additionally, it offers the benefits of fostering increased physical comfort within our bodies.

Again, we can look at how we tune the **precision** of these signals by adjusting the confidence and focus of our attention to shape our experiences and outcomes. For example, starting a creative project might stir up nervous energy or uncertainty. Instead of fixating on those sensations, someone could acknowledge them as natural and then redirect their focus to the first brushstroke on a canvas or the opening sentence of a story, imagining the flow and satisfaction of engaging in the creative process. As precision sharpens, internal signals come into line with external goals, and greater flow and clarity take shape. If our precision is too high, irrelevant signals (like chronic pain without a physical cause) can dominate our experience. If too low, critical cues (like thirst) might go unnoticed. Embodiment moderates this balance, holding emotional stability even as we respond to the constant shifts of our cycles.

Similar to learning to ride a bike, which combines sensory input, cognitive

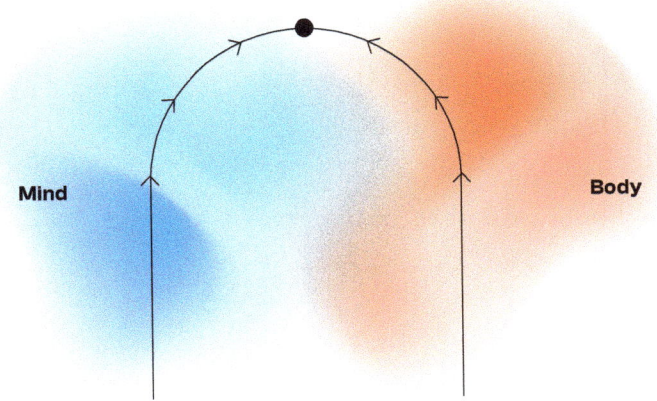

**Embodied solutions** often more deeply optimized for outcome potential

**Mind**

**Body**

know-how, and action, embodiment is best learned by doing. It's a practice best cultivated through action and lived experience, rather than abstract understanding. Admittedly, deep embodiment can be a challenging skill to learn. At present and in the context of this book, sparks provide easily available opportunities to facilitate deeper embodiment. There are numerous activities shown to foster embodiment, including yoga, meditation, somatic practices, breathwork, dance, and immersion in nature. This variability means we offer these activities as options for you to explore, encouraging you to tune into what resonates and adjust as needed. We've chosen to provide the best insights and practices available today, supported by active inference science and the larger body of literature.

One foundational hope I hold for the future is that greater accessibility of biofeedback tools will help us better understand and cultivate embodiment. Advances in technology, such as immersive virtual reality or sophisticated neural feedback tools, could provide new avenues for deepening embodiment more efficiently through simulated environments or enhanced cognitive training. However, the time, resources, and sometimes even the depth of personal challenge involved in making these shifts can limit accessibility for many. One of the deepest hopes I hold for humanity and the next forty years, is development in this sector and how it could help shift life experiences for a larger mass of our population. Not only by reducing unnecessary suffering and uncertainty for large swaths of people, but also by fostering more stable societies and environments that better support collective well-being.

Expanding access to biofeedback tools for embodiment presents meaningful challenges. Key design and engineering hurdles include reducing cost, improving usability, ensuring portability, and enhancing the reliability of indirect biomarker indicators. Tools must also work across diverse populations and be engaging enough for daily integration. Yet if we've been able to develop smart watches and rings that help us track stress and exercise (something I've had the privilege to work on in the past), there's reason to believe we can evolve these technologies to better support practices like meditation and embodied self-awareness. It's a space I hope to directly contribute to, making inner awareness as tech as accessible as physical fitness tracking has become.

## SPARK

One effective way to move from cognition into embodiment is through a simple body scan meditation. Find a comfortable position and slowly scan your attention from head to toe, gently noticing sensations without trying to change them. If you encounter areas of tension or discomfort, simply acknowledge them without judgment. Alternatively, you can also utilize guided body scan meditations through apps or streaming platforms. Taking even a brief moment for this practice can foster greater clarity, reduce uncertainty, and cultivate a more intuitive sense of embodied self-awareness.

# PLAY: REDUCING UNCERTAINTY THROUGH JOY

# PLAY: REDUCING UNCERTAINTY THROUGH JOY

Learning embodiment often involves encountering challenging sensations. One way to balance these experiences while continuing to update our active inference cycles and reduce free energy is through play. Play goes beyond just fun and games. It offers a way to explore the world, challenge our abilities, and discover our agency. From an active inference perspective, play serves a valuable purpose: resolving uncertainty about what we can and can't control. Joy can then be understood as our intrinsic signal that emerges when uncertainty is successfully minimized and predictions about the world align with sensory inputs in a meaningful and rewarding way.

## Tiny Scientists at Work

Take, for example, the simple joy of a baby kicking a mobile hanging above their crib. More than movement, it becomes an experiment. The baby moves their leg, observes the mobile sway, and stops—repeating this process in a carefully orchestrated way. What might look like flailing is actually a purposeful test of control. Through playful interactions, they uncover fundamental truths of their active inference cycles: I can make things happen.

This process of experimentation, observation, and learning builds the foundation of our world model, teaching us cause and effect long before we can articulate it. Uplifting experiences, such as joy and curiosity, is central to making this possible, reinforcing our learning experiences and motivating further exploration. This is especially crucial in early development, where children begin to grasp the concept of "self" and "other." Before we can understand ourselves, we must first recognize that others exist independently of us. The opportunity to learn through play is a gift we can readily offer the next generation. With space and opportunity, their development unfolds naturally.

## The Social Side of Play

Play often evolves from personal exploration to social interaction. From a baby discovering how to engage a parent through playful cries to children synchronizing their movements in a game, play opens the door to learning the rhythms of relationships. It offers space to explore the boundaries between self and other, sharpen a sense of agency, and engage with the subtleties of social interaction. The joy and safety found in these moments create conditions where exploration, expression, and resilience can naturally emerge. In this way, play becomes a setting for experimenting with how the world works and how we function within it through our active inference cycles.

Play doesn't lose its value after childhood. In adulthood, it offers a way to explore fresh ideas, experiment with different roles, and open up new possibilities. It interrupts rigid expectations and reconnects us with curiosity. In play, the mind opens, and we try out new ways of thinking and problem-solving without the weight of judgment. Moments of shared play help build trust and mutual understanding, reinforcing connection and collaboration. At the neurological level, play activates the brain's reward system, releasing chemicals that enhance learning, bonding, and motivation. Even in adulthood, it continues to be a powerful pathway for adaptability, resilience, and growth.

## Reconnecting with Play

If you're like me before writing this book, you might have forgotten how to play as an adult. Many of us stop playing, or perhaps didn't have the opportunity to practice much, due to a combination of cultural, psychological, and practical factors that gradually deprioritize play in our lives. Cultural expectations often prioritize productivity over leisure, psychological barriers such as fear of judgment or failure can create resistance, and practical challenges like demanding schedules and responsibilities can feel like they leave little time for play.

However, as we've continued to explore in this book, reducing our uncertainty (reducing our free energy) can be a deeply rewarding intention. Play, often in connection with others, offers us opportunities to do so in joyful ways. Your prior experiences and intentions mean that incorporating more play into your life will require taking steps that are uniquely yours.

At the bottom, you'll find an exercise that might inspire you to explore play in a way that feels authentic and enjoyable, should you wish to engage with it. I'll also offer my experience as inspiration. For me, it wasn't so much about learning how to play. Similar to still knowing how to ride a bike from childhood, I found those skills were still available to me. Rather, my own self-consciousness and propensity to judge myself limited my ability to step back into those familiar states of play.

As I aligned my intentions and understanding more closely with the realities I observed in the world, I began to realize that some of society's messages around achievement and external approval didn't fit. With this realization, judgments faded (both about myself and others) and I found that play naturally resurfaced. Incorporating it more with my friends, partner, and even dog has been a joyful shift in my life, and I've deeply appreciated the stronger bonds, trust, and resilience it has brought to those relationships.

When early environments lack opportunity for play, it often takes intention (and sometimes rewiring) to reopen the playground later in life. That's why play is now being used in therapeutic contexts, from video games that offer safe spaces for feedback and experimentation, to role-playing games like Dungeons & Dragons that help people explore emotional dynamics through story and imagination. These playful formats give people the chance to step into different roles, experiment with choices, and build new responses, all while having fun. **Play reminds us that growth and healing don't have to be heavy—it can be curious, creative, and even joyful.**

In essence, play is not a frivolous pastime but a joyful, essential part of how we learn, grow, and connect. In other words, play updates our active inference cycles and reduces free energy. Whether we are developing new skills, engaging in creative exploration, or strengthening relationships, play provides a powerful

" In essence, play is not a frivolous pastime but a joyful, essential part of how we learn, grow, and connect.

avenue for deeper self-discovery and a greater sense of agency in the world. I hope that this deeper understanding of play brings you increased joy and connection, both with yourself and in connection with others.

## SPARK

If you're curious to explore more play in your life, consider this gentle invitation: Can you remember a moment (recent or long ago) when you felt truly playful? What were you doing, and what did it feel like in your body or mood? Let that memory guide you. Is there something small you might try this week that carries the same spirit? It doesn't need to be perfect or planned—just a few minutes of lightness and play. What ended up feeling meaningful to you, and how might you invite more of those experiences into your days?

# A PLAYGROUND OF POSSIBILITIES: DISCOVERING A NEW PLACE WITHIN

# A PLAYGROUND OF POSSIBILITIES: DISCOVERING A NEW PLACE WITHIN

Over the course of writing this book, I discovered a new kind of exploration—one that felt different from the type of play we just examined. Rather than relying on jovial connection with others, a space emerged as a safe curiosity within myself. It was like stepping onto an internal playground: expansive, inviting, and safe with a variety of areas to explore, each offering its own unique sensations and learning opportunities. Like a child wandering between a climbing wall, a sandbox, and a swing set, each corner of inner space brought different sensations and self-knowing. My nervous system, once habitually on high alert, started offering a quieter baseline—an embodied signal that things were okay.

Before doing this work, I hadn't fully realized how weighed down I'd been by subtle cycles of unease and dissonance. As I practiced, interacted, and kindly explored these sensations, I began to feel more spaciousness within. More room to notice, shift, and engage, both internally and externally. This state was deeply personal, and I imagine others might describe it as spiritual, scientific, or something else entirely. But the essence is simple: as humans we can cultivate greater ease and contentment.

This transformation was deeply rooted in both the cognitive and felt aspects of my active inference cycle. As a researcher, I had long been accustomed to absorbing information and analyzing content, but I began dedicating similar attention to my body's felt sense. This intentional balance between knowledge and felt sense allowed me to cultivate a more accurate, adaptive understanding of myself and my surroundings.

Scientifically, this is a personalized example of experiencing the process of mentalizing interoception—becoming skilled at interpreting the signals the body sends and integrating them into your sense of self (similar to our discussions in Chapter 15). Without the capacity to explore these sensations, our internal signals can feel like noise—something to be ignored or endured. But by learning to pay attention and curiously explore them, I was equipping myself with valuable information to refine my model of the world.

Many of us grow up thinking there are only a few ways to care for or understand our bodies; perhaps through a mindfulness app or the occasional massage. But once I stepped into this internal playground, I realized just how many types of swings, slides, and corners there are to explore. This wasn't a one-size-fits-all space; it was a multidimensional terrain filled with different approaches, each revealing new insights about myself. Play fixtures such as meditation practices that went far beyond typical mindfulness apps, somatic and body-based work well outside mainstream yoga studios, group dynamics that extend past standard improv exercises, and cultural rituals that carry embodied wisdom from generations. Each offered new ways of sensing, knowing, and integrating. What mattered wasn't choosing the "right" practice, rather it was cultivating a spirit of curiosity and safety that let me explore and learn.

**ANNOTATION**

This might be a good time to acknowledge that, just as in any expansive space of exploration, not every path or practice is safe or grounded. Like a playground, it's wise to approach unfamiliar methods or individuals with thoughtfulness— especially those who make grand promises, diverge significantly from well-established science, or trigger certain types of unease. Cultivating discernment is part of the learning process, too.

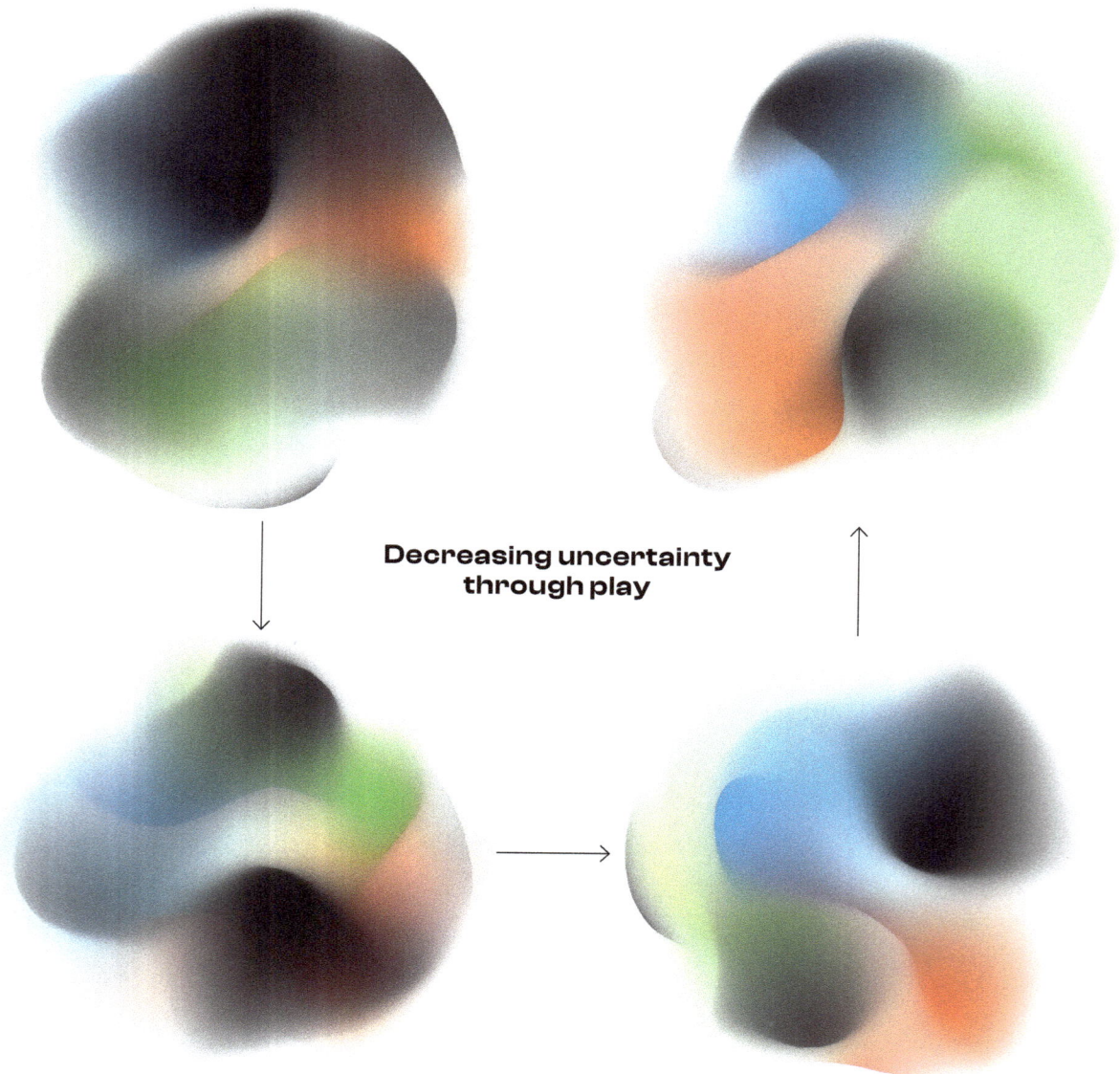

**Decreasing uncertainty through play**

**Navigating Uncertainty**

The shift came gradually and often with difficulty, with the focus on developing the tools and awareness needed to respond effectively to whatever arose. At first, I imagined reaching this place would feel like earning an honorary PhD, having put in the years of study and finally arriving at a kind of endpoint. But instead, it feels like I've just discovered a vast terrain still waiting to be explored. There's so much our systems are capable of, far beyond what many of us (or even current science) fully understand. That realization fills me with hope: that humanity holds untapped potential, and that we may be capable of things we've yet to even imagine.

If you find yourself in a similar place, navigating the tension between cognitive understanding and embodied experience, I encourage you to explore your own playground. Notice the signals your body is sending, navigate the uncertainties, and trust that through exploration and practices, clarity will emerge. After all, the joy of discovery lies not in having all the answers, but in learning to dance with the unknown.

## SPARK

- Consider trying a new practice that invites self-knowing through felt experience. This could be a variation of something familiar, such as a different style of yoga or a slower, more reflective approach to movement. Or it might be something entirely new, such as a local dance event or drum circle that connects you with rhythm and embodiment.

- What sensations do you notice as you begin? How does your body respond to novelty or stillness? What feels safe, unfamiliar, or alive in you?

# REST & THE GIFTS OF INTUITION

# REST & THE GIFTS OF INTUITION

Rest is often perceived as a passive state, a necessary pause in the midst of our busy lives. However, from the perspective of active inference and self-evidencing, rest plays a much more complex and essential role in how we navigate the world and refine our internal models. Here, the evidence for our mental models is sharpened, precision held alongside complexity, and the pause expanded into something fuller.

## The Science Behind Rest

We can consider rest at multiple levels. Whether it's sleep, periods of quiet introspection, or even brief moments of disengagement from sensory input, rest allows us to harmonize our understanding of the world. It helps strike a balance between two key components of self-evidencing: **accuracy** (how well our internal models fit our sensory perceptions) and **complexity**, which refers to the degree of detail in our models.

If our models become too complex, we risk overfitting and adding unnecessary details that hinder adaptability. For example, overanalyzing a simple decision, like choosing a meal or an outfit, can lead to decision fatigue rather than clarity. Similarly, micromanaging tasks by focusing on excessive details, such as

**DEFINITION**

Process by which our brains continuously seek to maximize the evidence for their internal models of the world. Essentially, the brain is constantly testing its predictions against sensory input to ensure that its understanding of the environment is as accurate and useful as possible.

dictating every step of a delegated project, can slow down progress and create inefficiencies.

Conversely, if our models are too simple, we risk underfitting and failing to capture important nuances of our experiences. For instance, making quick decisions without considering relevant factors, such as hastily choosing a meal or an outfit without reflecting on context, can lead to dissatisfaction. Likewise, delegating tasks without providing enough details or oversight can result in misunderstandings and incomplete work. Rest is a key method to help us arrive at this balance as we navigate life. It's like a form of cognitive housekeeping, allowing us to refine our models by simplifying them without losing essential information. As Einstein famously put it, we aim to *"keep everything as simple as possible, but no simpler."*

## Sleep as Neural Housekeeping

A clear example of the role of rest in optimizing complexity is the sleep-wake cycle. Throughout the day, we accumulate vast amounts of sensory data, forming countless associations and synaptic connections. Many of these are redundant or irrelevant in the long run. During sleep, our brain engages in a crucial process known as **synaptic homeostasis**, a concept proposed by Giulio Tononi. This process involves pruning unnecessary neural connections while also strengthening those most relevant to learning and memory, leaving behind a refined, simplified structure that is better suited for the challenges of the next day.

In essence, sleep restores energy while also simplifying and reinforcing our internal models so they stay efficient and adaptable. Without rest, our models risk becoming cluttered with excessive complexity, making it difficult to extract meaningful insights from our experiences.

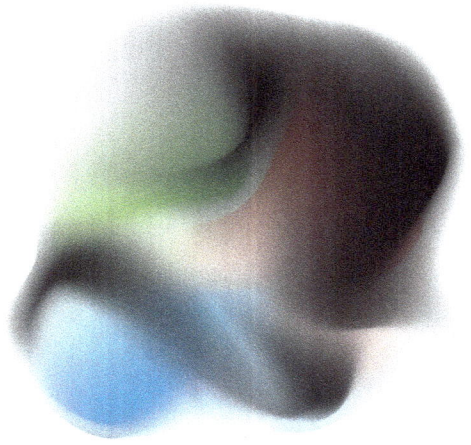 

**Limited sleep** and synaptic homeostasis

**Balanced sleep** and synaptic homeostasis

## Moment-to-Moment Rest

Beyond sleep, rest also occurs in shorter, more transient forms throughout the day. When we disengage from sensory input, such as closing our eyes for a moment, staring off into the distance, or engaging in quiet reflection, we create space to reorganize our thoughts and avoid cognitive overload. This momentary disengagement prevents the overfitting of our internal models by offering a chance to recalibrate and prioritize what is truly relevant. It's why we instinctively

**Navigating Uncertainty**

take breaks when processing complex information—our brain needs the space to simplify and consolidate.

## The Survival Perspective

From a survival standpoint, rest might seem counterintuitive. Amid the pressures of survival in an ever-shifting world, rest exposes us to potential risks. Yet it exists across species, suggesting its fundamental importance. Survival has favored organisms that can balance activity with rest, balancing their internal models to respond more effectively to their environments.

If survival were solely about constant vigilance and action, we wouldn't observe so many instances of rest across nature. Rest stands as a fundamental aspect of life, shaping the development of adaptive, well-tuned models that guide decisions and sustain resilience over time.

## Personal Reflections on Rest

I also like to think of how giving myself rest gives me gifts back. My body plays a key role in perceiving the world for me, and those senses are critical in refining my understanding of it (as discussed in Chapter 15). I find that when I give my body the gift of rest, it repays me through gifts of intuition. It's an entirely different way of working. Where before, I pushed through long days into long weeks, relentlessly running to achieve end goals, incorporating balance has shifted this working pattern. I'm still aware of strategic deadlines that increase my opportunities for impact; after all, if I were to write and research indefinitely without releasing a book, I wouldn't be able to sustain myself or continue my studies in this larger domain. Instead, I've found a new balance. One where I am committed to an early rise and sleep cycle because I know I'm most productive in the earlier hours of the day. This shift has also helped me strategically scope the content of this book. Most importantly, it has added greater depth, perspective, and value to the content that was ultimately included. Not only does it create a more meaningful offering, but it's also an experience I'm not just enduring—rather, it's an experience I want to live.

## Practical Implications

Understanding rest through the lens of active inference can shift our perspective on its importance in daily life. Rather than viewing rest as a passive or indulgent activity, we can see it as a powerful tool for cognitive balance. Incorporating intentional rest, whether through sleep, meditation, or moments of quiet reflection, can help us:

- Enhance learning and memory integration
- Improve decision-making and problem-solving
- Reduce mental fatigue and overwhelm
- Strengthen emotional resilience and well-being

In giving rest its due place, the mind regains clarity and the body finds room for more flexible responses to life's challenges. Incorporating rest into our daily routines is not a sign of weakness but a critical tool in our volitional toolset. Whether it's a good night's sleep, a conscious break, or a quiet walk, rest is an

**ANNOTATION**

Personally, I find this to be an effective strategy for staying relevant in a world with ever-increasing computational capacities. While computers can perform increasingly complex calculations, they remain limited in embodied sensemaking, refinement, and prompting—areas where our minds and bodies can excel, particularly when we allow ourselves adequate rest.

active process of simplifying and optimizing our internal models. It is in these moments of disengagement that we cultivate the clarity needed to face new challenges with a refined sense of agency.

So, the next time you find yourself overwhelmed by complexity, consider: rest is a vital way of navigating the art of life.

## SPARK

Is there one small change you feel called to make that could improve rest's effectiveness? It might be something simple, like using a light-based alarm clock in the winter, avoiding blue light before bed, or committing to a consistent wake-up time without hitting snooze. Perhaps it's limiting light leakage in your bedroom or trying a white noise machine. Or maybe it's something more foundational, like committing to longer, more consistent sleep each night.

# THE ARTIST WITHIN THE ACTIVE INFERENCE CYCLE

# THE ARTIST WITHIN THE ACTIVE INFERENCE CYCLE

We are living, breathing manifestations of an active inference process—a dynamic interplay of mind, body, and environment, working together in a rhythm that both shapes and is shaped by us. Yet, even though much of it occurs outside of our control, we are marvelously equipped for this process, and our physiology is finely tuned to guide us through life. However, within this automatic unfolding lies a transformative opportunity and a portion of the process where we can actively choose how to respond within. Now that we understand how active inference moves through us, we can now turn our focus more deeply to the parts we can transform.

In this chapter, we explore this interplay through metaphors and illustrations, reflecting on how we engage with physical discomfort. As I write this, I reflect on my own experience as a shifting, organic being, given the beautiful opportunity of life, yet bound by the constructs that enable and limit this opportunity. My body senses the delights of human touch, the warmth of sun rays on a spring day, and the signals of discomfort when I'm emotionally stressed or physically hurt. These sensations form the undertones of my existence, and as the artist of my life, I get to choose which colors and materials to engage with, when I engage versus observe, how much confidence to give them, and how much I trust the models I construct.

## Two Scenes, One Walk

Let's break this down. There are moments in my day when I feel joy, happiness, and contentment. For instance, most afternoons, after shutting my computer, my dog and I take a walk in the local park. I delight in the sound of his happy feet prancing at the announcement of "walk time." I take pleasure in the wag of his tail as he sniffs the park's finds and pulls us toward his discoveries. I'm calmed by the symphony of birds conversing in the trees, and in the spring, the lush overgrowth of trails, and the fresh, moist air coupled with the sun's warmth on my skin. These elements paint a scene of serenity within my active inference cycle—a moment of beauty consciously manifested.

But this experience, like all others, includes elements beyond my control. To reach the park, my dog and I must cross a busy intersection, navigating smelly, noisy cars on a narrow sidewalk. My dog, a little rescue with past trauma, feels the need to protect us from other dogs and people, oddly enough, wearing hats. These unpredictable elements create a complex triage of stressors, activating my body's systems as I maneuver through these challenges. It feels as though my limbic system (part of my larger threat and arousal systems) is extending its tentacles across my consciousness, heightening my senses and pulling my attention into a more aware state, readying itself to respond to what's happening around me. Writing about it now, I can feel my shoulders tense as my body responds to the undertones of this scene.

Ultimately, these two vastly different scenes—one of serenity, the other of stress—are part of the same walk to the park. Much of this experience exists

"As the artists of our active inference cycles, we are limited by our physiological palettes and environmental constraints. But within these limits lies creative potential.

outside my control: the size of the sidewalk, local traffic regulations, my dog's learned fears, and systems such as my limbic system's protective responses. This same system, though, contributed to saving our lives when a car sped through a crosswalk last year, enabling me to act swiftly and pull my dog and me to safety. These systems, such as the limbic system and its occasional discomfort, are integral to my organic existence. Without them, my dog and I wouldn't be making these walks at all.

## Curator of Action

However, as the curator of my active inference cycle, I can choose how I paint these scenes. For instance, I've learned to carry my dog during the most stressful portions of the walk, holding him close to my chest. In winter, he's often zipped into my jacket, with only his little head poking out, much to the neighborhood's amusement. This small adjustment transforms a potentially tense situation into one of warmth and connection. By doing so, I reduce the free energy associated with our walk, replacing stress with a sense of calm and security.

## Curator of Attention

Even with these adjustments, there are moments when tension creeps in. On particularly stressful days, I feel it pull into my neck and shoulders, sometimes tugging at an old injury. At this juncture, I face a decision tree. If I dwell on the sensation, it grows, coloring my experience with stress. My focus on the discomfort amplifies it, pulling me further into a cycle of tension and fear. But I've learned to recognize this pattern. Instead of spiraling, I shift my attention to the warmth of my dog against my chest. This isn't about denying the discomfort but about choosing where to focus. It's a careful dance of prioritization—believing that I can feel discomfort without it escalating, believing that my body's signals serve a purpose, and believing my ability to update my understandings over time.

## Sticky States

Many of us will experience chronic pain at some point. Pain, especially chronic pain, is a complex phenomenon influenced by environmental stressors, mental constructs, and physical sensations. Over time, pain can become less about an actual injury and more about a learned cycle. For instance, an old shoulder injury might heal, but the pain persists due to unresolved stress or habitual focus. This phenomenon, in part, can be explained by the brain's plasticity—the way it adapts and rewires itself based on repeated patterns. Pain, especially chronic pain, involves more than just physical damage; it becomes embedded in the brain's predictive coding system. The brain learns to associate certain sensations with danger or injury, even when the original cause has healed, perpetuating the experience of pain. This is a classic example of how an active inference cycle can get caught in a less effective loop, as the brain continually updates its model in ways that sustain rather than alleviate discomfort.

From an active inference perspective, these sensations can signal a need for recalibration, the need to make changes in our environment, or to seek further medical care. Chronic pain reflects intricate processes that can sometimes become stuck, rather than indicating failure. The brain's predictive coding system, which continuously updates models of the body and environment to reduce surprise, can at times reinforce unhelpful patterns. Researchers like Andy Clark highlight how cognitive constraints and environmental factors shape these processes, emphasizing that individuals are doing their best within the limits of their resources and context. This understanding reinforces the importance of viewing these challenges through a lens of compassion, recognizing that behavior and responses are shaped by the complex interplay of biological, cognitive, and environmental factors and not because someone isn't "trying hard enough".Turning attention from blame toward understanding fosters empathy and recognizes the intricacy of these neurobiological processes.

**Crucially, the brain's adaptability (and its plasticity) also offers hope, providing opportunities to recalibrate and rewire these cycles toward healing and growth.** We are fortunate to live in an era where advancements in neuroscience and psychology provide us with invaluable insights that previous generations lacked, offering new opportunities to understand and address painful cycles that can get stuck in our systems. If this is something you or someone you care about is working through, we've included resources at the end of this chapter. It's also valuable to remember that these experiences are not a reflection of weakness or personal failure. They are a testament and result of the complexity of how our minds and bodies adapt to challenges over time.

**ANNOTATION**

Similar to chronic pain, episodes of elevated heart rate—like those that occur during an anxiety attack—follow a similar dynamic. We've included resources for this as well at the end of the chapter.

## The Artist's Opportunity

As the artists of our active inference cycles, we are limited by our physiological palettes and environmental constraints. But within these limits lies creative potential. We get to choose which sensations to engage with, what to mentalize, how much confidence to assign them, and how to practice and learn. This agency allows us to craft experiences that contain more ease and harmony, even when discomfort arises. For my dog and me, this means our walks are largely serene and painted in colors of connection, peace, and exploration, even if they include moments of stress. From a broader perspective, this also paints the colors of my life experience. Many moments of peace, ease, connection, and creativity color my experience, even though it includes times of stress and discomfort.

## Resources

For Chronic Pain: Pain Reprocessing Therapy (PRT) is an emerging approach that teaches the brain to reinterpret pain signals. A 2021 randomized controlled trial in JAMA Psychiatry showed significant pain relief in chronic back pain patients using this method. While promising, PRT is still in its early development phase—more research is needed, and access may be limited. However, for some, it may offer a helpful path forward. Learn more at:

painreprocessingtherapy.com.

For Anxiety and Elevated Heart Rate: The Unwinding Anxiety App by Dr. Judson Brewer uses mindfulness and reward-based learning to help break the habit loops that drive anxiety and physiological overactivation. Backed by clinical research, it provides a practical and accessible tool for calming the nervous system and building awareness of body-based patterns. More at:

[unwindinganxiety.com](unwindinganxiety.com)

# LAYING OUT OUR PALETTE: GETTING TO KNOW OUR VOLITION

# LAYING OUT OUR PALETTE: GETTING TO KNOW OUR VOLITION

We've now explored the intricate systems that shape our active inference cycles and how they fundamentally and beautifully support us with impressive capacity, operating largely beyond our conscious awareness. We've also begun to recognize our role as artists, actively choosing how to color our experience by selecting which parts of the palette to engage with and how much attention to devote to each. Now, let's take a deeper look at the range of potential palettes before us and how they relate to active inference, the free energy principle, and the inevitable discomfort and dissonance woven into our beautifully complex constructs.

## Choosing Our Brushes: Four Possible Contrasts

### Inner Critic vs. Self-Compassion

The way we speak to ourselves shapes how we process experience. The inner critic metaphorically narrows our prediction landscape, reinforcing expectations of failure or inadequacy. Self-compassion, on the other hand, softens rigid predictions and allows for more adaptive learning. Active inference blossoms on updating models in response to new information; self-compassion creates a more effective space for this process by reducing defensive rigidity.

**PRACTICE:** Notice when critical thoughts or physical tensions start to build. How can you get curious and pick up a different brush? Is it pretending your inner voice is a good friend talking to you about the situation? Is it pausing and mentally naming it? Is it reframing, such as "mistakes are a natural part of learning"? Is it taking a moment for somatic self-compassion, such as deep breaths or putting your hand on your heart? How can you get creative with the plethora of paintbrushes before you respond to the uncomfortable dissonance and create more space for ease and creativity?

## Judging vs. Non-Judging Awareness

When we assign rigid labels to experiences, such as "bad," "wrong," "a failure," we reinforce rigid priors, which makes it harder to use prediction errors adaptively for recalibration. This extends beyond personal failures, such as judging others, situations, or even emotions as inherently good or bad can similarly lock us into rigid models of reality. We might judge sadness as something to avoid, discomfort as inherently negative, or another person's actions as purely wrong without considering the broader context. Active inference is enhanced when we engage with experience without immediate judgment, allowing our models to update organically and making space for more nuanced understanding and adaptability.

**PRACTICE:** In moments of challenge, explore shifting from judgment to observation. Perhaps instead of "*I failed,*" something like "*I'm experiencing difficulty, and learning is a process.*" Notice if you're assigning a rigid label to the moment—are you calling an emotion '*bad*,' assuming someone's intent, or labeling an experience as '*wrong*'? Maybe broadening your perspective by asking, *"What else might be happening here?"* or *"What might I be missing?"* is a useful brush to utilize. Or perhaps taking a moment to label the sensation or judgement starts to make a significant shift for you. Just as we explored in the inner critic and self-compassion practices, the key here is to experiment and discover what approaches best support you in shifting your experience. Each small shift contributes to the evolving artwork of your active inference cycle.

## Rigid Control vs. Adaptive Navigation

When we grasp for excessive control, we resist the natural unfolding of experience, failing to reduce free energy effectively by fighting against uncertainty. Adaptive navigation accepts that life is unpredictable while still providing a sense of direction, helping us move more effectively through its constant shifts. Rather than rigidly clinging to a specific outcome, we can practice setting a broad intention and adjusting our approach in response to real-time feedback. Just as a sailor adjusts their sails to the wind rather than trying to force the ocean into compliance, we, too, can work with circumstances rather than against them. Along with other methods, these practices support active inference in flowing fluidly, easing tension and promoting comfort in uncertain situations.

**PRACTICE:** Instead of focusing on rigidly controlling outcomes, explore centering your attention on your intentions. Ask yourself, *"What's my guiding principle in this moment?"* and allow flexibility in how you navigate toward it.

**ANNOTATION**

I've found it helpful to cultivate a small positive feeling when I notice the opportunity to update and learn. For example, during meditation, when I realize I've drifted, I gently evoke a sense of kindness—framing it as a positive chance to practice coming back.

Consider shifting from a fixed expectation of results to an adaptive approach that embraces course corrections as part of the process. You might reflect on past experiences where an unexpected turn led to positive growth.

Another way to cultivate adaptive navigation is to experiment with small, low-stakes decisions: try letting go of minor preferences or experimenting with a different way of doing something routine. Observe how it feels to embrace fluidity over rigidity and note any shifts in ease or creativity. Once again, we've laid out some possible paintbrushes. What are some you might want to explore? What others might you want to fashion and use?

### Reactive vs. Responsive Engagement

Reactivity often arises from deeply ingrained priors as automatic responses that may no longer serve us. It might manifest as an immediate defensive response, a tendency toward people-pleasing, or catastrophizing situations. These reactions often reinforce outdated predictive models, increasing rigidity in how we engage with the world. Responsiveness, by contrast, invites presence and intentionality, allowing for more precise updating of predictions. By creating space between stimulus and response, we make room for a more adaptive and intentional way of engaging with our experiences. Instead of reacting defensively, one might practice active listening, acknowledging emotions while responding with curiosity rather than resistance. For people-pleasing tendencies, responsiveness could look like pausing before saying *'yes'* and asking, *'Does this align with my values and energy?'* rather than automatically accommodating others. In moments of catastrophizing, responsiveness might involve grounding techniques, such as focusing on present sensory details or reframing thoughts by asking, *"What is the most likely outcome rather than the worst possible one?"*

> **PRACTICE:** When faced with an emotional trigger, you could practice slowing down before reacting. Take a slow, intentional breath, and notice the physical sensations in your body, creating space between the stimulus and response. If defensiveness arises with others, you could try practicing active listening by acknowledging emotions and responding with curiosity rather than resistance. If the urge to people-please emerges, perhaps pause and ask yourself, *"Am I saying 'yes' out of alignment with my values or energy?"* If catastrophizing takes hold, perhaps reframing the situation, asking, *"What is the most likely outcome rather than the worst possible one?"* You might also find it helpful to visualize a more measured response or to remind yourself that emotions are temporary. Each of these small shifts helps cultivate responsiveness, which directly supports our process of active inference manifestation.

You can see that even these four core principles we've drafted to help provide context and examples overlap with each other. In reality, it's not four rigid constructs; rather, it's a beautifully interwoven palette of opportunity and one that you get to shape and utilize in your own way. Consciously choosing which elements to engage with shapes life experiences that resonate more deeply, harmonizing with uncertainty and aligning perspectives with a flexible, adaptive reality.

## Expanding Our Palette

We could fill this book with ways to expand the nuances of this palette, and in some ways, we largely have. For those of you who find it helpful, we've briefly listed several additional "artistic tools" to color your active inference manifestation below.

### Inaction vs Action

Inaction is a form of action; rest and recovery help maintain physiological and mental balance by activating the parasympathetic nervous system and supporting homeostasis. Just like we can't inhale forever, we need pauses to recalibrate and regain clarity. Without this rhythm, constant doing narrows perception, while too much stillness may lead to disengagement and model entropy.

> **PRACTICE:** Notice moments when you default to either constant action or prolonged inaction. If you find yourself always pushing forward, try pausing for intentional rest by taking actions, such as taking a walk, lying down, or focusing on your breath. If inaction is your habitual response, experiment with a small, deliberate action, such as reaching out to someone or completing one simple task. Observe how these shifts affect your sense of balance and ease.

### Contraction vs. Expansion

Stress and fear tend to shrink our perceptual field, limiting possible actions by heightening sympathetic arousal and narrowing attention to perceived threats.

Expansion through openness, curiosity, and presence activates broader networks in the brain, allowing for more flexible predictions and adaptive action in response to new possibilities. Under sympathetic arousal, our mental field narrows like a tunnel, but with safety and curiosity, it can widen like a landscape coming into view.

> **PRACTICE:** When noticing contraction, consciously engage in something that broadens awareness through physical movement, deep breathing, or simply shifting attention outward to one's surroundings.

### Sensory Avoidance vs. Sensory Engagement

Avoiding discomfort can reinforce maladaptive predictions by signaling to the brain that the sensation is dangerous, keeping us stuck in avoidance loops. Engaging with sensation (even discomfort) in a measured way allows for recalibration by updating the brain's predictions based on new, safe experiences. The brain, like a skittish animal, can update through calm exposure that the once-feared path is now safe to tread.

> **PRACTICE:** Perhaps try gradually leaning into uncomfortable sensations, holding space for them without immediately reacting or escaping.

### Cognitive Rigidity vs. Playful Exploration

Fixating on rigid interpretations of experience limits adaptability by sustaining prediction errors and leaving them unresolved. A playful, exploratory stance, like treating experience as an experiment rather than a test, supports the brain's capacity to revise its models in response to new, less threatening input.

> **PRACTICE:** Some of these constructs might feel like a brain teaser. Which ones might be fun to explore? What are some ways you could increase play or jest in this exploration?

## Forcing Resolution vs. Allowing Integration

Seeking an immediate resolution to discomfort can prevent deeper learning by reinforcing the brain's urgency to eliminate uncertainty rather than understand it. Letting experiences integrate over time allows for a more sustainable recalibration, as new neural pathways form through gradual meaning-making and prediction updating. Just as a muscle strengthens through rest between repetitions, integration can benefit from spaciousness, not constant effort.

> **PRACTICE:** Instead of rushing to fix discomfort, experiment with small shifts, explore different responses, and allow space for when an intended shift doesn't happen. Observe how the feeling evolves, much like discovering a new approach unfolding over time.

## Boundary Setting vs. Boundary Diffusion

Healthy boundaries help define the conditions for sustainable, adaptive inference by regulating the flow of sensory and emotional input. It's similar to echoing everyone else's voice so often that your own becomes inaudible; self-knowing is limited when internal signals are constantly overridden.

> **PRACTICE:** Consider reflecting on where you tend to overextend. Perhaps practice small, clear boundary-setting statements, such as *"I can't commit to*

*that right now, but I appreciate the offer."*

## Blame vs. Ownership

Blame externalizes control, reinforcing a passive stance in one's predictive land-scape. Ownership or acknowledgement reclaims agency, allowing for active participation in model updating. The shift supports both personal growth and healthier connections, allowing for mutual understanding and collaboration instead of reinforcing cycles of defensiveness or resentment.

> **PRACTICE:** Shift from *"This happened to me"* or *"You did this"* to *"How do I want to respond to this?"* or, from a neutral perspective, *"What happened here?"* Ownership and acknowledgement do not mean self-blame but rather recognizing one's role in navigating forward.

## Rumination vs. Reframing

Rumination, cycling through the same thoughts again and again, keeps us stuck by reinforcing existing predictions rather than allowing for flexible adaptation. Reframing, on the other hand, helps introduce new perspectives, making space for predictive recalibration and adaptive learning. Rumination is like replaying the same scene in a movie, hoping the ending will change, while reframing edits the script to explore a different resolution.

> **PRACTICE:** Perhaps reflect on if you've lingered on this thought several times in the past, leading to the same outcome. Or writing down recurring thoughts and then questioning their accuracy or imagining alternative viewpoints could be helpful.

## Becoming Skilled Artists

As artists within our own experiences, we are all still learning. No artwork is perfect, nor can it be. There are times when we paint with less skill, when our strokes are uneven or rushed, or the volitional opportunity to shift the tone, such as working our way through the grief process, is quite limited. But the beauty of the palette is that we have agency. Every moment presents an opportunity to reach for a new brush, to shift how we engage with the colors of our experience. Even as I write this, I, too, still find myself painting with old habits at times. And that's okay. Because the masterpiece of life isn't about flawless execution—it's about engaging, learning from strokes, and continuously shaping a composition that reflects more ease, connection, and coherence. What matters most is that we keep creating.

## SPARK

As we've likely begun to notice, much of the active inference cycle operates beneath our awareness, and given that this kind of internal learning is rarely modeled or reinforced in our broader society, it can be helpful to find small habits that support it. One such habit is engaging with media (such as podcasts or books) that gently reinforce these principles. However, it's important to choose content that aligns with your values and fosters genuine integration rather than overwhelm or spiritual bypassing. I can't tell you what that looks like for you, but I can share what worked in part for me: a blend of mindfulness teachings and science taught by someone open to diverse spiritual perspectives without imposing a particular belief system.

# THE VALUE OF LOVE: A GUIDE FOR UNDERSTANDING & CONNECTION

# THE VALUE OF LOVE:
# A GUIDE FOR UNDERSTANDING & CONNECTION

It's one thing to refine how we respond to the world. It's another to realize how much of our world is illuminated through connection with others. Love is not just an emotion or an abstract ideal; it is one of the most profound organizing principles of human life. Across cultures, religions, and philosophies, love has been regarded as a central force in morality, a fundamental source of meaning, and a key contributor to fulfillment. Christianity speaks of *agape*, an unconditional love that transcends the self, while Judaism emphasizes *ahavah*, a love rooted in covenant, justice, and ethical responsibility, unconditional love that transcends the self. Buddhism teaches *metta*, or loving-kindness, as a practice for dissolving barriers between oneself and others. Hinduism embraces *bhakti*, devotional love, as a path to the divine. In Islam, love is at the heart of one's relationship with God and the community. These traditions, though diverse, all point to a fundamental construct—love is essential for human flourishing. It offers meaning, connection, and a lens through which we understand both ourselves and the world.

The word '*love*' in English carries a vast range of meanings, often requiring additional context to fully capture its depth. In contrast, many languages have multiple words for different forms of love. Greek, for instance, distinguishes between *eros* (romantic love), *philia* (friendship), and *agape* (universal, unconditional love). In English, we use the same word to describe our feelings for a romantic partner, a close friend, a favorite food, or a personal passion. This limitation can make it difficult to articulate the many ways we experience love in daily life. When we describe love in this chapter, we include the quiet gratitude we feel for a collaborator whose dedication and insight inspire us. Or the warmth that arises when we see a small child laughing, even if we have no personal connection to them.

Love is also present in the appreciation we hold for a skilled practitioner, such as a musician, a doctor, or a therapist, who has spent years honing their craft and shares their expertise with generosity and care. These experiences may not fit neatly into traditional notions of love, yet they evoke a sense of connection, reverence, and mutual recognition. Understanding love in this broader sense allows us to appreciate its many forms and how it relates to our individual and collective active inference cycles.

## Love as an Inference Process

If we think about love from an active inference perspective, we can see it's fundamental to our manifestation of our cycles because it expands our sense of self not only physically but cognitively as well. This happens as our brain integrates the other person into both our bodily regulation and our internal model of the world. Many of us are already quite familiar with the feeling of love as presented by contemporary society. Here, we have an opportunity to pause and consider how active inference might reshape that understanding, revealing how love, through processes of bodily regulation and predictive modeling, deepens both our perception and experience of connection. Whether romantic, familial, platonic, or altruistic, cognitively it acts as a <mark>sense-making process,</mark> shaping how we engage with the world and with one another. It fosters prosocial behaviors, including emotional resilience, strengthens social bonds, and enhances our capacity for joy. Without it, we risk falling into isolation and rigidity, unable to see beyond the limits of our own perceptions.

**DEFINITION**

The process by which people give meaning to their collective experiences.

## Romantic Love

Romantic love, when viewed through the lens of active inference, is not merely an experience but a deep, embodied inference. It includes subtle cues, such as tones, gestures, and postures that our body often picks up before our conscious mind does. When we detect these subtle cues, they ignite a shift in our internal state; our heart rate adjusts, our posture softens, our nervous system responds. And in that physiological shift, our brain begins to make meaning.

In these moments, we begin to infer that another person is like us and they are, in some way, part of us. Not because we've run the logic, but because our interoceptive system, the part of us that interprets bodily signals, has generated a new prior: this is someone I belong with. Our preferences feel mirrored, our values echoed, and our self-model (our internal story of who we are) feels less like a solo act and more like a duet.

Romantic love, then, is a layered construction. It's rooted in the body's response to another, but it grows through the ongoing integration of shared narratives, mutual investment, and open feedback loops. **The science points to romantic love not as a mystery we fall into—but a meaning we build, piece by piece, signal by signal, together.**

## Friendships

Love, as a generative or mental model process, integrates bodily sensations and social interactions, forming our ability to connect and build relationships. When we engage with others, we update these generative models. I'm often reminded that many learning models suggest this happens most deeply after we come to know ourselves. Then deepen that learning through connection with others. These updates happen through sensory cues updating our internal predictions to align more closely with reality. In romantic relationships, the feel of a partner's embrace or the subtle shift in their tone provides meaningful feedback.

In friendships, a knowing glance or shared laughter reinforces connection. Even in professional relationships, a mentor's encouraging nod or a colleague's thoughtful feedback shapes our expectations and deepens trust. Across these various bonds, sensory cues function as essential signals that continuously update our understanding of others and enhance our emotional attunement. This ongoing process allows us to foster trust, anticipate needs, and create a shared reality with those we love.

Building on this, love moves beyond individual awareness and unfolds as a continuous cycle of mutual inference. When two people care about each other,

**Love and
Active Inference**

**Mind**
Sense-making, curiosity,
cognitive resonance, compassion,
mental updating

**Environment**
Mutual attunement, relational
updating, shared narratives,
empathic signaling, mutual inference

**Body**
Embodied inference, emotional
regulation, physiological attunement,
sensory cues, intentional engagement

they actively adjust their internal models in response to one another's actions, emotions, and expressions. In this ongoing exchange, they come to read each other more accurately, emotional attunement takes root, and their bond holds with greater strength. Over time, this dynamic process ensures that love remains flexible, responsive, and resilient—an ever-evolving force that sustains connection and meaning in our lives.

When I love someone, I can also foster that relationship by becoming more attuned to the warmth and connection that love brings. This heightened awareness influences my actions, encouraging me to nurture the relationship by seeking closeness, asking thoughtful questions, or engaging in small gestures of affection, like a reassuring touch or an embrace. Love, in this way, also actively influences how we move through and engage with the world around us. Remaining attuned to the warmth and connection means actively choosing to take in new perspectives, remain open to change, and allow our internal models of others to update. This process is essential for love to flourish, as it enables us to see others more fully, adjust our assumptions, and build trust through shared understanding.

## Self-Love

Self-love, from an active inference perspective, is the ongoing process of updating our beliefs about ourselves with kindness and curiosity. Just as we learn to see others more clearly by remaining open to new evidence, we can learn to see ourselves more accurately—not through judgment, but through gentle inquiry. When we attend to our internal signals with compassion, we regulate our physiology, reduce uncertainty, and build a more stable, trustworthy internal model.

Practically, this means that self-love isn't about indulgence or ego—it's about learning to interpret our needs, emotions, and bodily cues in ways that help us stay grounded, safe, and open to growth. It increases our ability to regulate stress, make healthier decisions, and remain resilient in relationships. Without self-love, we are more prone to rigid beliefs, reactive behaviors, and chronic states of physiological or emotional dysregulation. Self-love, then, forms a foundation of emotional homeostasis that allows connection with others to take shape.

## Love as an Expansion of the Self

As love deepens, the boundaries between self and other blur and integrate. Individuals in close relationships increasingly integrate their loved ones into their predictive models, leading to a greater sense of shared identity. Rather than remaining isolated, love draws us into an expanded, more interconnected way of being. This ability to see oneself as part of a larger system aligns with broader ideas of social cohesion and altruism, but it is also inherently self-serving. Love enhances emotional resilience, reduces stress, and improves overall well-being. It provides a reliable support system, fosters personal growth, and helps individuals develop a more adaptable, flexible mindset. Even for those who are not motivated by altruism, love offers practical benefits: increased cooperation, greater efficiency in collaboration, and an enriched sense of purpose.

## The Role of Sensory Attenuation in Love

We now understand that through the lens of active inference, to love someone is to continually receive and integrate evidence that they are like us—that they share our values, intentions, or emotional landscape. This doesn't mean they're identical to us. It means there's enough overlap in our internal models of the world to sustain a shared narrative. But this inference depends on a simple yet essential condition: we must remain open to the signals that reveal this overlap.

When that openness disappears, when we begin to attenuate or filter out signals that someone might be like us, our predictions start to harden. Bit by bit, they become "not like me." Sometimes this means we find ourselves in relationships where we've formed a fixed perception of someone, and no matter how they show up, or who they truly are, we struggle to see them without distortion. This can result in a lost opportunity to connect and to benefit from the emotional resilience and joy that connection can bring.

At the same time, this can be balanced with the reality that not everyone wants or is able to align with our generative models—and there are moments when protective distance is necessary for our well-being. For me, with people I've needed to place kind and protective space between, I keep a gentle antenna tuned in that direction. I listen for signals that they may have shifted, or that they might now be in a place to meet me in a way that aligns, at least functionally, with my generative model. For instance, by respecting basic boundaries or acknowledging past hurts.

The process of *'not like me'* becomes even more concerning when it moves

---

**ANNOTATION**

I also find it really helpful to have genuine well wishes for them. This is scientifically useful because it supports emotional regulation, reduces defensive responses, and helps maintain cognitive flexibility. From an active inference perspective, holding positive intent keeps our priors from becoming overly rigid and allows us to remain open to new evidence—preserving the possibility of future connection.

from simple distance to <mark>dehumanization.</mark> When dehumanization happens we stop engaging, stop updating, and over time, stop perceiving. The other person becomes a fixed idea, not a living being. This process happens at the interpersonal level, and tragically, at the collective level too—in conflicts, ideologies, and wars where whole groups become invisible to one another. Left unchecked, this can lead to isolation, hardened biases, diminished empathy, and even the justification of harm. Thus, dehumanization is both morally problematic and strategically unwise. We trade genuine connection for control, but end up with less flexibility, more stress, fragile group dynamics, increased conflict, and less social protection. In the long run, dehumanization confines others, and in doing so, confines us as well.

The danger is that this behavior can feel rational. If you're certain that the other side will never change, it may seem "optimal" to retreat into your own camp and stop listening altogether. But active inference tells us something deeper: when we stop gathering new evidence, we cut ourselves off from the very signals that could revise our beliefs. Ultimately, we become trapped when error goes unquestioned and curiosity to re-examine our beliefs is absent.

Love, in this view, becomes a kind of epistemic bravery or vulnerability. It requires us to keep the channels open, even when it would be safer to shut down. **This doesn't mean staying in harmful relationships or exposing ourselves to repeated patterns of harm. It means keeping awake to the difference between someone who truly can't meet us and the possibility that we've lost sight of them or that they have changed.**

**DEFINITION**

The process of depriving a person or group of positive human qualities.

**Navigating Uncertainty**

" As love deepens, the boundaries between self and other blur and integrate. Individuals in close relationships increasingly integrate their loved ones into their predictive models, leading to a greater sense of shared identity.

Finally, we can see how sensory attenuation can also be beneficial in love. Long-term partners or close friends often unconsciously filter out minor irritations (like a partner's habitual way of telling a story or a friend's small quirks) so they don't get bogged down by unnecessary frustrations. This allows them to focus on the deeper emotional connection rather than being distracted by trivial details, and make room for appreciation, patience, and understanding, so the relationship endures with strength and resilience over time. Ultimately, this is yet another instance where volition is a valuable skill, discerning which signals are valuable to engage with and which are best left aside.

## Cultivating Love in Our Lives

We're starting to understand that love, in active inference terms, is the brain's inference that you are like me. It's a shift in belief that recognizes shared meaning, fostering trust and connection to emerge.

Yet this kind of recognition is far from automatic. As we've seen, rigid prior beliefs about someone can prevent us from updating in light of new evidence about who they are. The same applies when past experiences shape our expectations with others in the present, causing us to overweight familiar outcomes and overlook evidence that doesn't fit. These filters, often held below awareness, can quietly close the door to real connection.

To love, then, we can practice loosening the grip of those less effective assumptions. This might be a kind gesture, a shift in tone, a shared silence that feels unexpectedly safe. It's also possible to use the art palette techniques we discussed in the previous chapter, but in social contexts. We can explore practices like forgiveness, reprocessing past pain, or revisiting old assumptions to further reduce the weight of outdated predictions that may otherwise distort our social perception. This frees up bandwidth in our generative models, allowing more accurate and compassionate inferences about others, and ultimately creates more space for authentic connection in the present.

Seen this way, love is the act of relaxing mental rigidity in favor of curiosity and the opportunity to shift. It's an invitation to update what we believe about one another, not blindly, but compassionately. It's a process that allows people to reduce uncertainty together by co-creating a shared space of meaning. And in doing so, it reminds us that love is not the absence of uncertainty, but a willingness to explore it together.

Practices like meditation are another tool we can use to foster deeper love in ourselves and with others. It enhances our ability to recognize and regulate our internal states, improving our capacity to connect. For those who struggle with emotional attunement or social connection, this ability can be transformative. Through greater attentional control and awareness of interoceptive signals, individuals can become more attuned to their bodily cues, such as heart rate changes, breath patterns, or muscle tension. These often serve as unconscious indicators of emotional states. Interpreting and regulating these signals expands our capacity to both give and receive love more deeply. Similarly, conscious engagement with our physiological responses to love cultivates self-awareness and helps us steward our relationships with greater clarity, trust, and emotional balance.

Understanding love as an active process means recognizing that it requires intentional engagement by practicing attuned connection with each other in various forms. Love is not something that simply happens; it is something we cultivate. By practicing curiosity about others, developing attentional control, and embracing the uncertainties inherent in relationships, we can deepen our capacity to love and be loved. The ability to love also utilizes the willingness to remain in a state of exploration, adjusting our perspectives as our relationships grow and shift. If love were merely a fixed state, it would quickly stagnate. The process of love is, in many ways, an ongoing commitment to learning and updating when we remain fluid and responsive to the complexities of life.

At its core, love is an invitation to expand, to connect, and to embrace the beauty of uncertainty. In a world that can often feel chaotic and fragmented, love is a key path forward.

## SPARK

- What comes up if you pause to reflect on your relationships? Is there a relationship that might be calling your attention you've been ignoring? What arises when you give it attention? Is there an action you're feeling pulled towards?

- Love doesn't always bloom through grand gestures or perfect timing. Love can grow through small moments of listening to what's alive inside you, and sharing that in a way that brings more understanding, not more harm.

" Love moves beyond individual awareness and unfolds as a continuous cycle of mutual inference. When two people care about each other, they actively adjust their internal models in response to one another's actions, emotions, and expressions.

# DEEPENING SELF-KNOWLEDGE: UNDERSTANDING THE MIND'S LAYERS

# DEEPENING SELF-KNOWLEDGE: UNDERSTANDING THE MIND'S LAYERS

Having just explored love through our relationships with others, we now turn inward to consider how cultivating safety and care within ourselves can deepen our self-knowing. Throughout this book, we've touched on how increasing our self-knowing enables more effective active inference cycles and reduces our free energy and causes an internal sense of dissonance when our predictions about the world and our experiences don't align. Now, in the sections below, we will dive deeper into the expanded supportive process through historical approaches to consciousness combined with modern neuroscience. However, instead of getting lost in nuanced debates, we will focus on broad examples that highlight why this self-knowing can be useful.

## Self-Love as a Gateway for Self-Knowing

In active inference, safety lowers the cost of uncertainty, making greater exploration possible. When we treat ourselves with care and compassion, our system registers that sense of safety. This internal signal lowers the biological and cognitive cost of confronting uncertainty, making us more open to revising our inner models. By contrast, harsh self-judgment reinforces rigid priors—fixed predictions that limit adaptability. Self-love eases this rigidity, allowing emotional truths and evolving preferences to be integrated more flexibly. Active inference also depends on reading our internal signals clearly. Self-love increases our opportunity to feel what's truly present, such as grief, vulnerability, or longing, without suppressing or distorting it. In this way, self-love provides a neurobiological foundation for deep, honest self-knowing, going beyond mere comfort.

These kinds of shifts can feel incredibly difficult, especially when inner harshness has helped us survive. I noticed this in myself recently when I examined the judgment I held toward the soft curve of my belly. Like many women in my family, I tend to carry some weight there. The world around me has often reinforced the message that this part of me is undesirable, suggesting I should change my eating and exercise habits to fix it.

For years, I carried a quiet dislike for this area of my body. When someone touched it, I would instinctively pull away. More subtly, that inner tension affected how I stood, how I moved, and how I saw myself. It held a quiet, dissociative energy. As I deepened my own journey of self-knowing, I began to realize this discomfort wasn't just about how I looked. It was interrupting the places in my body where I might otherwise feel confidence or even love.

This realization helped me craft a gentle exploration that aligned with the science. I intentionally offered kindness to that part of myself when it came into my awareness. I also started reframing my thoughts, and feeling gratitude for the organs in my abdomen that work continuously to keep me alive, allow me to function, and do things I enjoy.

What began as an experiment has led to real change. I feel more confident.

I can access more of a sense of love in my chest that used to feel blocked. And I feel the relief of no longer carrying that quiet burden. Perhaps unsurprisingly, my belly hasn't grown since releasing that judgment. Instead, I often have a bit more energy and that extra energy helps me make choices that are healthier and more connected.

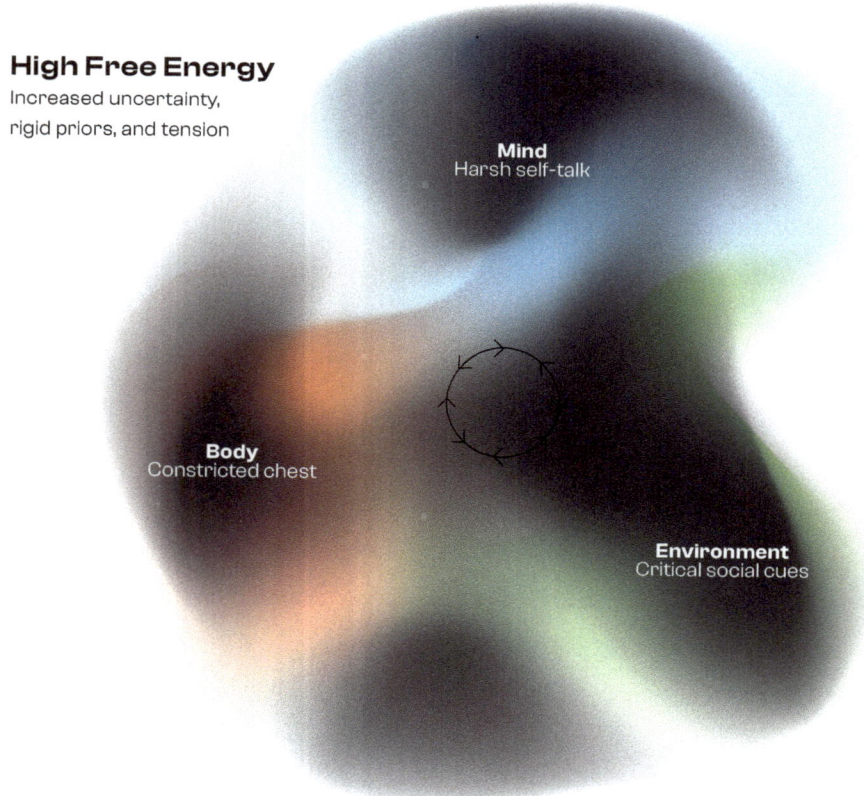

**High Free Energy**
Increased uncertainty,
rigid priors, and tension

**Mind**
Harsh self-talk

**Body**
Constricted chest

**Environment**
Critical social cues

Our systems are wired to protect us when they sense danger. When we approach ourselves with criticism, those systems respond accordingly by pulling away, locking up, or bracing against change. But when we approach ourselves with care, we lower that inner resistance and open space for self-knowing to emerge. Letting go of self-criticism and cultivating greater compassion is a powerful way to reconnect with ourselves and discover who we truly are.

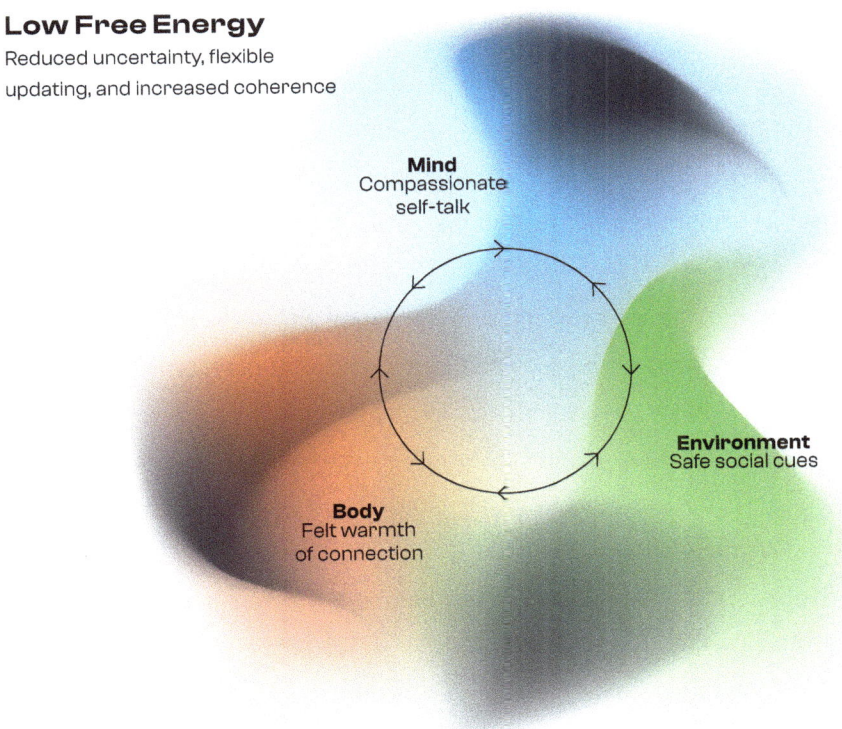

**Low Free Energy**
Reduced uncertainty, flexible updating, and increased coherence

**Mind**
Compassionate self-talk

**Environment**
Safe social cues

**Body**
Felt warmth of connection

## The Layers of the Mind

Having established a sense of safety and self-compassion, we're now better equipped to explore the science of our minds. Throughout history, many traditions have described different levels of consciousness. Freud's model of the mind introduced the primary process, associated with instinctual, emotion-driven thought (the id), and the secondary process, which is logical, structured, and tied to our sense of self (the ego). One way to read Freud's distinction, in light of modern neuroscience, is as a hierarchical inference machine between these two processes. Meaning the mind continuously generates predictions, evaluates sensory input, and updates its internal model to minimize the difference between the two. Drawing on Freud's ideas, this model links to a wider body of psychological and neuroscientific research that supports the distinction between these two cognitive processes.

The primary process aligns with id-driven thinking, characterized by free-associative, imaginative, and emotion-driven cognition. It is the language of dreams, deep instincts, and spontaneous creativity. It operates at a more fundamental, deeper, and more instinctual layer of our consciousness, influencing our thoughts and behaviors in ways that are often outside of our direct awareness.

The secondary process, or ego-driven thinking, introduces structure, suppressing chaotic impulses and refining thought into a coherent narrative aligned with external reality. It operates at a higher, more deliberate level of our consciousness, guiding our reasoning, decision-making, and social interactions in ways that shape our everyday experience. These two processes are constantly

**ANNOTATION**

This discussion draws on Freud's widely accepted contributions to understanding the mind, rather than some of his more controversial theories.

active, running in parallel and exchanging information beneath our conscious awareness. While we may not always notice it, the interplay between them shapes our perceptions, emotional responses, and decision-making. This ongoing interaction ensures that we can access creativity, imagination, problem-solving, emotional intelligence, decision-making clarity, adaptability, social intuition, impulse regulation, abstract thinking, self-awareness, and a stable, organized sense of self.

The coordination between the primary and secondary processes is one place where the Free Energy Principle comes in. Our brain constantly works to minimize the uncertainty (or free energy) between what we expect and what we experience between these two levels. When the brain successfully integrates sensory, emotional, and cognitive signals, we feel aligned, stable, and present. When it fails, we can experience internal conflict, anxiety, or a sense of detachment.

A key player in this balancing act is the Default Mode Network (DMN), a large-scale brain network that engages in self-referential thought, memory, and identity formation. Once again, we find ourselves in a space where subconscious processes, in this case the DMN, quietly shape our perceptions, emotions, and decisions. These largely invisible mechanisms fuel a highly capable active inference cycle, continuously working in the background to refine our understanding of the world. And, as we'll see later on, we have an opportunity to bring these processes into greater awareness, allowing us to consciously influence and shift the way we experience and interpret our reality.

## When Regulation Fails: The Challenges of Misalignment

The DMN plays a central role in integrating the primary and secondary processes, allowing us to maintain a coherent self while navigating the world. However, when this system is disrupted, it can lead to dream-like thinking in waking life, which manifests as intrusive thoughts, flashbacks, or dissociation. This disruption can stem from various sources, including chronic stress, extended sleep deprivation, unresolved trauma, substance abuse, poor diet and nutrition, social isolation, and overuse of digital media. Each of these factors can interfere with the brain's ability to integrate sensory and cognitive signals, making it harder to maintain a stable, coherent self-experience.

If the brain cannot effectively suppress chaotic signals from lower levels, it leads to heightened free energy and, consequently, greater stress and dissonance. We may feel overwhelmed by emotional loops, struggle with rumination, or become disconnected from reality. To compensate for this misalignment, the mind and body expend additional resources trying to stabilize perception, often at the cost of our well-being.

For example, suppressing unresolved trauma requires continuous effort to keep distressing memories from surfacing, leading to chronic tension and heightened emotional reactivity. Substance abuse artificially alters perception, forcing the brain to work harder to reconcile distorted signals with reality, which can result in cognitive fatigue and dependency. Overuse of digital media overstimulates attentional networks, disrupting the brain's ability to regulate focus and emotional processing, leading to heightened stress and diminished resilience.

Over time, these compensatory efforts deplete mental and physical energy, exacerbating stress, anxiety, and emotional dysregulation.

However, once again, we are truly fortunate to be among the first generations with access to robust science that helps us understand and shift these experiences. This means we have the ability to make meaningful changes not only in our own lives but also in the patterns and cycles that shape future generations. Even small shifts can have profound ripple effects, creating a new foundation for ourselves and those who come after us. While we do not need to resolve everything, the progress we make today lays the foundation for a future with greater mental clarity, emotional resilience, and a deeper sense of connection, both within ourselves and with those around us.

The takeaway? When our primary and secondary processes are misaligned, our system works harder to make sense of the world, draining energy and reducing our ability to engage meaningfully with life. The more we cultivate self-knowing and refine our active inference cycles, the more we reduce this burden, freeing up cognitive and emotional resources to support deeper engagement with ourselves and others. Understanding ourselves in this way helps us respond more skillfully to what's happening inside us. It also contributes to shifting long-standing generational patterns, creating more space for clarity and well-being in future generations.

## Some Practical Tools: Meditation, Dream Journaling, and Therapeutic Approaches

One way to bridge these gaps is by engaging with our unconscious mind in structured ways. Meditation, for example, strengthens our ability to observe thoughts without becoming entangled in them. This practice cultivates meta-awareness, helps us notice when deeper process thinking arises, and offers a chance to regain stability. As we tune into bodily sensations and the transient nature of thoughts, meditation creates space for noticing without immediately acting on them in our higher level processes. That shift in awareness helps us recalibrate our interpretation of experiences, preventing unconscious patterns from dominating our perception and behavior.

Dream journaling is another tool that offers insight into unconscious thought patterns. Recording and reflecting on our dreams is an effective process in which we can surface hidden emotions, recognize patterns, and integrate aspects of our psyche that may otherwise remain buried. However, balance is key. Dreams are best understood as symbolic, rather than taken literally, as they reflect the mind's attempts to reconcile uncertainty rather than direct representations of reality.

Additionally, psychoanalysis and therapeutic approaches rooted in understanding unconscious processes can help us navigate internal conflicts and provide us with a structured way to integrate the disparate layers of our mind. These methods offer valuable tools for identifying patterns of thought and behavior that may be operating outside our conscious awareness, helping us recognize and reframe limiting beliefs. Seeking these resources can take many forms, from working with a therapist trained in depth psychology to exploring practices such as Internal Family Systems (IFS) or somatic therapies. When we bring curiosity and authentic openness to these practices, they allow for deeper integration, and they strengthen our emotional resilience and psychological flexibility.

In reality, we all experience a certain level of misalignment in our systems—it's an inherent part of the beautiful opportunity we've been given in life. Outside of debates about the existence of states like sainthood, there is no final destination where misalignment is entirely removed  Instead, we can view it as an ongoing opportunity to explore, refine, and align ourselves as we move through life. And if you're like me, you may find it helpful to explore multiple avenues to help ease dissonance in various forms, creating an additive effect that allows us to move through life with greater ease and fulfilment.

## A Window into Psychedelic Therapies

While meditation and various forms of introspection can be powerful tools, some individuals choose to seek additional pathways for deepening self-awareness. Psychedelic-assisted therapy is a promising and fast developing field, in part because it can temporarily dissolve the rigid structures of secondary processes like the default mode network and allow for greater integration of suppressed thoughts and emotions.

It's important to note that psychedelics are not required for self-exploration, but they can be a helpful asset when approached with the right context and preparation. Functionally, psychedelics appear to reduce hierarchical suppression in the brain, particularly within the DMN, allowing previously constrained information to surface. This temporary loosening of rigid thought patterns enables novel perspectives and deeper emotional processing, which can lead to significant shifts in self-perception and integration of past experiences. This can create an opportunity for deep emotional processing and facilitate a more flexible, adaptive model of the self.

From an active inference perspective, psychedelics temporarily increase entropy (or free energy) in the system before helping it settle into a more stable, updated configuration. This can allow individuals to work through entrenched fears, traumas, and cognitive patterns that no longer serve them. However, the effectiveness of this process depends on the set and setting and the mental framework, environment, and support structure in which the experience occurs.

Our understanding of foundational science points to the potential of this resource in reducing free energy, the dissonance and uncertainty within our own active inference cycles. At this point, however, medical systems are primarily structured to rigorously evaluate their impact on those struggling the most. For example, research has shown particularly promising results for PTSD, treatment-resistant depression, and existential distress at the end of life, with studies highlighting significant reductions in symptoms for veterans and others facing deep psychological pain.

However, emerging evidence continues to suggest benefits beyond clinical diagnoses, indicating that psychedelic-assisted therapy likely also supports

**Navigating Uncertainty**

"When we're more attuned to ourselves, we become more predictable to others and to ourselves.

individuals seeking deeper personal insight and emotional resilience. This field is developing quickly, and it is still in the early stages compared to long-established interventions like cognitive behavioral therapy, with limited long-term studies and extensive meta-analyses available. Nonetheless, while the science is still emerging, these therapies hold real potential for meaningful change. When used properly, they can facilitate deep shifts for those who choose to engage with them, offering a rare opening for self-exploration and healing. However, it remains essential to approach these therapies in the larger context; the true transformation occurs in the integration process that follows. Without proper support, dosage, and context, an altered state can be destabilizing rather than healing.

## When Others Choose a Different Path

If you're reading this book, you're likely one of the many who have decided to embark on life in this way. To cultivate a lifestyle with less dissonance and free energy by getting in touch with deeper parts of yourself. However, not everyone makes this choice. For some, it's an unconscious decision—they never knew otherwise. For others, the temporary discomfort outweighs the long-term benefits. Coming to terms with this reality is often difficult to accept. It can be initially unsettling and painful, but over time, acknowledging this reality ultimately reduces our free energy and allows for more stable, meaningful connections with those who have consciously made more skillful decisions.

For me, the hardest part wasn't just recognizing that some close connections choose not to engage with painful signals or make meaningful changes.

Accepting this truth was both difficult and important. The next challenge was figuring out how to honor my own path while respecting theirs, for example, how to express a need without evoking an aggravated response. I'm still exploring this balance. However, when I am able to navigate this balance, it enables relationships to exist within the overlap, rather than feeling an even greater void from severing the connection entirely.

If we think of the selfish-unselfish dance from earlier and the need to maintain our own homeostasis, we can apply a few key guideposts to support these opportunities. A primary one is through boundaries. This term is often overapplied in common vernacular, but in this instance, it refers to ensuring that those we choose to connect with, regardless of their pursuit of reduced free energy, respect our need to maintain our own balance of homeostasis. Various models exist in the literature, such as Brene Brown's BRAVING framework, which provides structure for establishing and maintaining boundaries. At their core, this model helps us create enough space for others to live as they choose while still preserving our ability to follow our own path and benefit from the meaningful rewards of connection.

Painfully, this also requires accepting when someone's level of free energy (and the outward ripples of pain within their system) renders them unable to respect our needs. Recognizing the limits of our own resilience, navigating the mix of relief and grief that comes with separation, and finding clarity in when and how to step away are difficult yet necessary resolutions. These choices often require careful reflection and a willingness to integrate the complexity of our interwoven experiences.

---

**ANNOTATION**

BRAVING consists of seven key elements: Boundaries (setting clear expectations), Reliability (following through on commitments), Accountability (owning mistakes and making amends), Vault (keeping private information confidential), Integrity (choosing courage over comfort), Nonjudgment (allowing for open, honest conversations), and Generosity (assuming the best in others).

However, even when others take a different path, we still have the opportunity to offer quiet support. When we reduce our own free energy and move toward greater coherence, that internal shift often extends outward. This touches those around us and shares the benefits of our process in subtle, meaningful ways.

## Self-Knowing and the Shape of Loneliness

Loneliness takes shape when uncertainty lingers, even in the presence or absence of others. Uncertain whether we belong, whether we're valued, whether we're truly seen by others. That uncertainty shows up in our physiology, in the background tension of our system, and in the stories we quietly tell ourselves about who we are and how others might receive us.

When we don't know ourselves well, those stories tend to default to old protective patterns: withdrawal, defensiveness, over-efforting. We anticipate rejection, so we pull away. We feel unseen, so we stop reaching out. And because these reactions often keep us from gathering new evidence, our beliefs remain unchallenged. The loop of loneliness continues.

Self-knowing begins to shift this. When we understand our internal signals, such as our anxiety, our need for closeness, and our sensitivity to disconnection, we become less afraid of them. We gain the ability to pause, reflect, and gently ask: *"What's really happening here?"* That moment of pause is powerful. It gives us more choice in how we respond, and more softness toward the parts of us that hurt.

And it changes how we connect.

When we're more attuned to ourselves, we become more predictable to others and to ourselves. We communicate more clearly. We become easier to read, easier to trust. Relationships rely on mutual understanding, and that's much harder to build when we're operating from self-doubt or disconnection from our own needs.

The more we know ourselves (not perfectly, but compassionately), the more space we create for connection. Loneliness doesn't evaporate all at once, but it becomes less defining. Less sharp. Because with self-knowing, we're no longer waiting to be seen. We're learning how to see ourselves and how to stay open enough to let others in.

## A Quick Note on Epigenetics

Before closing, it's worth briefly touching on epigenetics. This growing area of science explores how life experiences and environments can influence gene expression without altering the underlying DNA. While it's a fascinating concept that underscores how we are shaped by both biology and experience, the field is still sorting out exactly how influential these mechanisms are and under what conditions they matter most.

In the context of this book, we want to acknowledge that while epigenetics does play a role in shaping our biology, it is likely not the central driver of our lived experience. Science continues to uncover the complex ways that environmental influences can affect gene expression, but many of these processes are context-dependent and still not fully understood. What we do know is that our environments, relationships, behaviors, and interpretations play a large and immediate role in shaping how we feel and function day to day.

Personally, I like to visualize epigenetics as one thread within a much larger tapestry—the bodily layer of influence within the broader active inference cycle we've been exploring. I can feel it, and even honor its influence, but I can also sense how much space remains for interaction, training, and adaptation. Just as a regular practice like yoga can reshape physiology and attention, our environments and habits can influence patterns of gene expression. To me, epigenetics feels like an influence that's present, but it doesn't define the space. However, awareness, intentional action, and thoughtful design of our surroundings can support shifts toward greater coherence and adaptability.

Interestingly, though perhaps unsurprisingly, our beliefs about epigenetics likely also matter. The bidirectional nature of the brain means that if we believe we are powerless against our biology, that belief itself can become self-fulfilling. But if we understand that we have influence, whether through awareness, intention, and meaningful shifts across our mind, bodies, and environments, that perception can offer a powerful route for change.

## Closing Thoughts: Choosing Our Path Forward

Self-knowing is a lifelong process. Whether we explore it through meditation, reflective practices, therapy, or altered states, the goal is the same: to reduce unnecessary free energy and bring greater coherence to our lived experience. When the gaps in our misalignment shrink, it becomes easier to feel balanced,

**ANNOTATION**

Even within therapy specifically, thoughtful and guided cross-modality use can deepen its effectiveness.

flexible, and connected with ourselves, those around us, and our environment. The more we refine this balance, the more we open ourselves to the exquisiteness of life, stepping into life with greater clarity emotional strength, and a richer understanding of what it means to be human.

## SPARK

Choose a part of yourself (physical, emotional, or behavioral) that you often judge or try to fix. Instead of pushing it away, experiment with offering it kindness. What happens in your body as you try this? Does anything soften or resist? Consider writing a short note of appreciation to this part of you, acknowledging the ways it may have protected you or helped you survive.

# SAFE CONTAINERS: THE FOUNDATION FOR GROWTH & CONNECTION

# SAFE CONTAINERS: THE FOUNDATION FOR GROWTH & CONNECTION

## What Are Safe Containers?

Self-knowing doesn't unfold in a vacuum; it requires space. A space where we feel safe enough to notice, explore, and revise our internal models. **A safe container is any environment (physical, emotional, or relational) that provides stability, predictability, and a sense of security within our active inference cycles.** These environments serve as essential foundations for neurodevelopment and shape how individuals may engage with uncertainty and belief updating. It allows us to explore, learn, and interact with the world without excessive fear or risk. They do not eliminate all uncertainty but offer a structured space where belief updating, self-discovery, and meaningful connection can occur with minimized distress and reduced free energy.

## The Benefits of Cultivating Safe Containers with Others

The benefits of cultivating safe containers are likely something you already intuitively understand, even if you haven't named it as such. For example, think about how you feel in relationships that offer steadiness compared to those that are unpredictable. Often, the difference you sense isn't just about behavior; it's about the underlying level of safety each relationship provides as a container for emotional connection and growth. Consider a father-son relationship where a child feels secure sharing intimate struggles. The son can express emotions freely, knowing he will be heard and supported. The father, having worked through his own anxieties and updated his belief cycle, minimizes his own free energy and fosters a sense of security, allowing his son to navigate his emotions with confidence. Contrast this with a relationship where the father constantly projects his own anxieties onto his son, assuming the same fears and limitations apply, or becomes overprotective, limiting the child's ability to develop confidence in his own decision-making.

The son, in turn, may withhold his thoughts, creating a distance between them. This does not mean either of them is "deserving" of suffering. The father may have had a difficult relationship with his own father and lacked access to the insights and resources available today. Likewise, the son is not trying to hurt his father by withdrawing but is instead instinctively protecting his own sense of homeostasis. Both are doing their best within the understandings they have inherited, and their ability to adapt and grow is directly linked to the lack of a safe container where belief updating struggles to occur within excessive distress. This distinction directly impacts the active inference cycle and free energy minimization.

This aligns with the idea that insecure or disorganized attachment styles act as 'unsafe containers,' where belief updating is constrained due to mistrust and the perceived instability of relational cues. This creates a cycle of misalignment and tension, preventing meaningful connection and mutual growth. The father may feel frustration, guilt, or helplessness when his attempts to guide his son are met with resistance or withdrawal. Without a sense of security, the son may internalize anxiety, become hesitant to express himself, and develop a felt sense of unease in his relationships.

Secure attachments, by contrast, create stable, predictable conditions that enable smooth belief updating. In a stable and supportive relationship, the child experiences predictable interactions, allowing for effective belief updating with minimal uncertainty. Likewise, the father benefits from this stability, as reducing

## High Free Energy Transfer

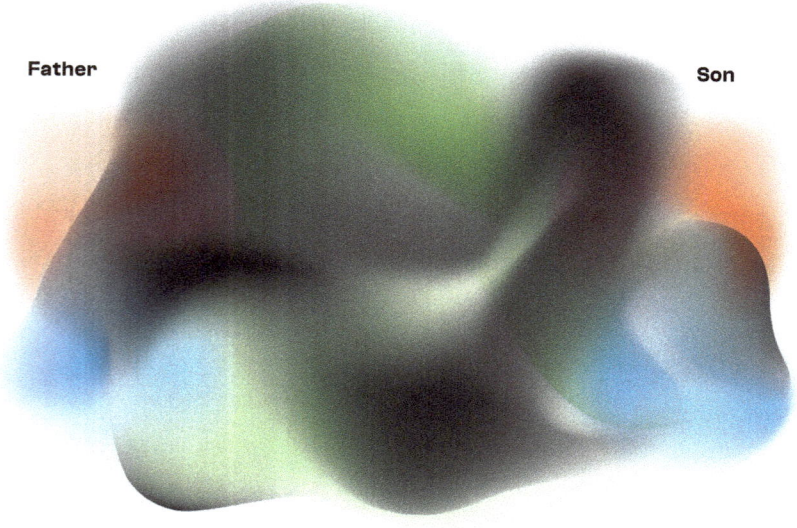

Father                                                                 Son

his own free energy allows him to engage in more constructive and emotionally regulated interactions. This cultivates a felt sense of ease and reduces chronic stress often resulting in the creation of space for more joy, fulfillment, and confidence in his role as a parent. For the son, this stability increases a sense of security and trust, enabling him the opportunity to navigate his emotions with greater resilience and self-assurance. Feeling seen and supported can allow him to explore his own beliefs and experiences without fear of rejection, fostering a deeper sense of self-awareness and emotional well-being.

## Low Free Energy Transfer

Father

Son

The same principle applies in the workplace. Consider a manager who is out of touch with her own need for stability and regulation. This state is what we might call "Regulatory Blindness." Within her own team, she is able to provide clarity and consistency, ensuring a stable environment. However, when interacting with other teams or external stakeholders, her internal instability manifests as unpredictable communication, difficulty in aligning expectations, or an inability to maintain a cohesive vision. Without realizing it, she creates uncertainty across departments, leading to breakdowns in collaboration and increased friction between teams. Over time, this misalignment contributes to confusion, inefficiencies, and growing frustration, as other teams struggle to integrate her directives within their own workflows.

**DEFINITION**

A state where someone appears steady in one area but misses how their actions create instability in another. In Active Inference terms, it reflects misjudging which signals to trust, so regulation works locally but fails more broadly.

Conversely, a manager who is attuned to her own regulatory needs and actively stabilizes her responses can extend that clarity beyond her immediate team, fostering smoother interdepartmental collaboration. This allows for more adaptive problem-solving, trust-building across teams, and greater overall organizational coherence. The presence or absence of a safe container at this broader level not only influences immediate team dynamics but also determines how successfully an organization as a whole can function and innovate.

Safe containers are beneficial to our experiences and extend their impact to those we interact with. Internal regulation and awareness reduce tension within ourselves and make it easier for those around us to find balance, too. Stability in these moments also supports clearer communication, stronger collaboration, and more flexibility when circumstances are uncertain. In parenting, leadership, or everyday social interactions, maintaining a sense of safety allows beliefs and understanding to shift naturally, fostering shared growth. Over time, this practice nurtures resilience, trust, and a deeper connection both inwardly and outwardly.

## Safe Containers at Different Scales

We've already explored examples of safe containers in connection with others, such as the father-son relationship and the workplace example. Safe containers exist at other levels. On an individual level, a person who engages in self-regulation creates an internal safe container, allowing them to navigate challenges with greater resilience. However, when less effective strategies like self-criticism dominate, as discussed in Chapter 22, this internal space can become compromised. Someone who persistently critiques themselves may generate excess free energy, making it difficult to process new information effectively. Instead of fluidly updating their understanding of their experiences, they may become stuck in cycles of limiting beliefs and behaviors. Cultivating a balanced inner dialogue increases the stability of our internal container and supports clearer beliefs that facilitate emotional regulation. This stability improves personal well-being and makes interactions with others more constructive and grounded.

At the national scale, systems such as healthcare, education, and governance function as macro-level safe containers, helping to manage uncertainty for citizens by providing predictable structures and reliable access to basic needs. This allows people to engage more productively in life rather than operating from chronic fear or instability. Just as an individual benefits from internal regulation, a stable society also benefits from being aware of how its structures contribute to or erode collective stability. Active Inference frames these systems as niche construction, scaffolds that reduce surprise by sustaining predictability. When a nation invests in policies that increase stability and access to trusted resources, it fosters an environment where individuals can maintain their own stability with greater ease. Conversely, unstable funding models and a lack of transparency stretch collective resources and undermine predictability, raising uncertainty at the systemic level. A well-functioning national approach ensures balance across competing needs, such as budget. Finding these balances reinforces the stability of the system as a whole, creating a feedback loop where collective and personal homeostasis support one another, reducing overall free energy and enabling more effective adaptation to our ever-changing environments across different scales.

We can take this thinking a step further and consider safe containers on a global scale. Just as individuals balance self-protection with the benefits of relational harmony, nations are required to navigate the tension between securing their own interests and contributing to broader stability. Investing too heavily in self-preservation at the expense of collaboration can lead to isolation, resulting in economic stagnation, social unrest, or even increased threats, as external entities respond with increased hostility or exclusion.

On the other hand, overextending without safeguards can create vulnerabilities such as erosion of core stability, dependence on unsustainable strategies, and unsustainable resource depletion. Once again, we find the need for balance and safety, this time at the global scale. This book does not aim to dictate global policy; however, understanding these principles can encourage us to reflect on and refine our perspectives to contribute to more effective decision-making.

## Somatic Awareness and Self-Regulation

As discussed in Chapter 15, somatic awareness, the ability to sense and interpret bodily signals, is central in interoception and emotional regulation. Developing this skill and bringing it into connection with others contributes significantly to the overall creation and maintenance of a safe container.

From the perspective of active inference, our brains are constantly making predictions about our internal state and the external world. When we engage in somatic awareness, we refine our ability to interpret these signals accurately, reducing uncertainty and minimizing unnecessary free energy. For example, noticing clenched fists or jaw tension before responding to criticism, then taking a moment to relax the muscles and refine the response before speaking calmly, rather than reacting impulsively.

Self-regulation is also valuable in maintaining a safe container, as it prevents the spread of dysregulation within social interactions. Taking a deep breath before responding, grounding oneself in physical sensations, or using movement-based strategies (such as shifting posture or softening tension in the body) can help stabilize internal states. This reduces personal distress while also helping lower collective free energy, making interactions feel more predictable and emotionally safe.

In group settings, self-regulation helps prevent emotional contagion, where dysregulated states spread and escalate uncertainty. **When one person models calm, grounded presence, it provides an anchor for others to do the same.** The natural ability of our systems to co-regulate and influence one another is a promising and hopeful dynamic I see in humanity. This response implies that collectively our responses are nonlinear. Sometimes it only takes one to incite a collective shift.

## Communication: One-Way Safe Containers Are Reinforced or Broken

One of the most crucial elements of a safe container is communication. This includes how we express ourselves, how we listen, and how we hold space for others. Communication is a vital yet constrained link between highly complex active inference cycles, enabling them to update and benefit from one another. Due to the limited nature of communications, we can benefit from being intentional about its use, optimizing it to reduce both personal and systemic free energy. Effective communication, through the lens of active inference and free energy, serves as a mechanism for reducing uncertainty and fostering alignment between individuals. It involves clear, honest, and non-reactive dialogue, minimizing unnecessary cognitive and emotional friction. When we listen to understand rather than just respond, adjustments in understanding happen more naturally, making social interactions less prone to surprises. This also allows us to engage with differing perspectives curiously, rather than defensively, which supports a stable exchange of information. In short, by communicating in ways that reinforce emotional safety, individuals maintain lower levels of free energy, allowing for more adaptive and cohesive social interactions.

Conversely, a safe container can be degraded by less effective communication, such as dismissiveness, judgment, or invalidation of another's experience. From the perspective of active inference, these communication styles less effectively use the limited cable of communication that connects us, increasing free energy while making it harder for both individuals to update their understandings and find cohesiveness. Reactivity and emotionally charged responses create instability by amplifying uncertainty, while attempts to manipulate or control the conversation prevent open exploration of new, potentially more effective perspectives.

Without a safe container for adjusting perspectives, communication can become rigid, reinforcing prior biases rather than fostering adaptation and mutual understanding. Imagine sharing a vulnerable thought with a friend. A supportive response might be, *"That sounds really tough. I'm here to listen."* A container-breaking response would be, *"You're overreacting. Just get over it."* The former creates safety, while the latter introduces instability and discourages further openness.

A great example of this is a sensitive, critical conversation. Imagine two partners discussing a difficult topic. They take the time to establish a safe container by ensuring they are both emotionally regulated before engaging, setting a shared intention for the conversation that creates safety for them both, and actively listening to one another without defensiveness. They acknowledge their own

# "By communicating in ways that reinforce emotional safety, individuals maintain lower levels of free energy, allowing for more adaptive and cohesive social interactions.

perspectives while remaining open to updating their understanding. As tensions arise, they use clear and non-reactive dialogue to navigate differences, reducing uncertainty and fostering trust. This creates an environment where both partners feel safe to express themselves honestly, leading to a constructive resolution with less dissonance remaining between them.

Conversely, when a safe container is not established, the conversation is more likely to derail. If one partner enters the discussion without a feeling of safety, reactive emotions are more likely to take over, leading to defensive responses and increased free energy. This escalates tension and reinforces uncertainty, creating a greater likelihood that both individuals retreat into less helpful, rigid perspectives. The lack of a stable communication environment prevents belief updating, leaving both partners feeling unheard and disconnected.

## Touch: The Subtle Power of Physical Presence

Touch, when appropriate and welcomed, is a profound way of reinforcing safe containers, both physiologically and emotionally. If verbal communication, actions, and somatic awareness are small yet powerful conduits that connect two systems, touch serves as another essential pathway, offering a unique and profound means of reinforcing connection and emotional safety. Supportive touch might look like a gentle hand on the shoulder to offer reassurance, a warm hug between close friends or family members, or physical gestures of care, such as holding a child's hand, a comforting pat on the back, or a tender caress between partners that conveys intimacy and trust.

In consenting contexts, supportive touch can help shift autonomic balance (often associated with lower stress responses and increased trust), which can ease belief updating and reinforce a sense of safety in relationships. This physiological regulation minimizes uncertainty and enhances belief updating, reinforcing a sense of safety within relationships.

That being said, a safe container can also be broken when touch is unwanted, forced, or used without consent. Similarly, an absence of any comforting touch in situations where it might be expected or desired can also create a sense of emotional distance. But while a grieving friend may find comfort in a reassuring touch, someone who has experienced trauma may feel unsafe with unexpected contact. Being attuned to the needs and comfort of the other person ensures that touch remains a reinforcing, rather than destabilizing, element of safety.

## Actions: How We Create or Undermine Safe Containers

Actions, both small and large, reinforce or weaken the integrity of a safe container. Supportive actions include following through on commitments to build trust, respecting personal boundaries, offering help when someone is struggling while allowing them the autonomy to accept or decline, and regulating our own nervous system before engaging in emotionally charged situations.

On the other hand, inconsistency, such as saying one thing and doing another, breaks trust, as does disregarding or crossing personal boundaries. Engaging in controlling behaviors, even with good intentions, can also compromise the safety of the container. Consider a therapy session where a therapist encourages vulnerability, assuring clients they can share openly. However, if the therapist frequently interrupts, dismisses emotions, or invalidates feelings, clients may begin to feel unsafe and hesitant to express themselves. This breakdown in trust increases free energy in the interaction, making it difficult for clients to update their beliefs about their experiences in a constructive way. Without a stable environment, clients may reinforce defensive behaviors rather than explore new insights. In contrast, a therapist who actively listens, responds with empathy, and allows space for emotions to be fully processed fosters a safe container. This encourages belief updating, reduces uncertainty, and supports deeper personal growth through active inference.

## Safe Containers and Neurological Development

Attachment theory highlights how early relational experiences shape neurodevelopment, influencing how individuals process uncertainty and regulate their emotions. Providing secure attachments for children, such as attuning to and responding to their needs, creates stable and predictable environments, reinforcing a sense of safety that enables belief updating and cognitive flexibility.

For example, a child whose caregiver consistently responds to distress with warmth and reassurance is more likely to develop a sense of trust in others, fostering resilience in future relationships. In contrast, insecure attachments, providing inconsistent support, or responding unpredictably to a child's emotional needs, contribute to heightened uncertainty. This increases free energy in the system, reinforcing defensive behaviors that make it difficult to integrate new information.

In another example, a child who experiences frequent invalidation or neglect may develop hyper-vigilance or rigid coping strategies, making belief updating and adaptation more challenging. These early experiences determine whether individuals develop epistemic trust (the confidence to accept and integrate knowledge from others) or epistemic mistrust, leading to skepticism and rigidity in thought patterns.

Jeremy Holmes, drawing from the Free Energy Principle, highlights that epistemic trust is essential for learning, adaptation, and psychological resilience. Insecure attachments disrupt this process, causing individuals to resist updating their beliefs even when new, corrective information is available. **This is particularly relevant in therapy, education, and leadership, where the presence or absence of a safe container determines whether individuals feel supported enough to explore and integrate new perspectives.**

We are incredibly fortunate to live in a time where research in neuroscience, psychology, and attachment theory provides us with the tools to shift these cycles—no matter where we find ourselves within them. Unlike previous generations, we have access to insights that can help us move beyond inherited patterns of uncertainty, mistrust, and emotional dysregulation. This makes us among the first to have the opportunity to consciously engage in reshaping these patterns in a meaningful way. If this feels like it resonates, you may find the additional resources below helpful. While these steps take effort, the rewards are substantial. Additionally, by actively participating in breaking these cycles, we not only improve our own well-being but also contribute to a larger environment of secure relationships and emotional resilience.

## Balance in Creating Safe Containers

While the benefits of safe containers are clear, they are most effective when maintained in balance—both for the individual and the group. A safe container is limiting when it becomes a rigid bubble that insulates us from all uncertainty or requires self-sacrifice to maintain. True stability arises from a mutualistic approach, where all individuals involved maintain their own ability to regulate and adapt. Growth requires some degree of challenge, but it should not come at the cost of one's own homeostasis.

If a safe container becomes too restrictive, it stifles growth and limits exposure to new ideas or experiences in our active inference cycles, creating a sense of confinement rather than support. Or if maintaining the container demands excessive emotional labor or the abandonment of personal boundaries, it can lead to exhaustion and imbalance. Thus, the most beneficial safe containers are co-created, ensuring that all individuals remain engaged in a way that supports their own and each other's well-being. This balance is fundamental at personal, relational, and societal levels, allowing individuals to navigate uncertainty while remaining grounded in the opportunity to self-stabilize.

## The Role of Epistemic Trust in Safe Containers

While we've touched on this indirectly, pausing to highlight the value of epistemic trust (the willingness to consider new knowledge as trustworthy) can be helpful, as it plays a significant yet often overlooked role in shaping how we learn and grow through relationships. Epistemic trust develops in stable, secure relationships where individuals feel recognized and understood. When safe containers are established through trust-building behaviors such as consistency, attunement, and validation, individuals feel secure enough to engage in belief updating. Conversely, when interactions are inconsistent, dismissive, or controlling, epistemic trust is weakened, making learning and psychological growth more difficult. Unkind treatment hampers both sides: the person targeted and the one trying to assert control, since restricting others in this way also limits their own opportunity to grow.

## The Limitation of Trying to Change Others

A common pitfall in relationships and communities is the tendency to try to change others to meet our needs. This is a fragile and unsustainable strategy because it places control outside of ourselves and imposes expectations on others. Instead of fostering connection, it breeds frustration, resentment, and instability.

A more sustainable approach is to focus on what we can offer, such as a safe, regulated presence that invites connection and curiosity without forcing it. One gift we can offer is our own regulated system, which naturally provides a sense of stability for others. When we engage with others from a state of balance, we create conditions where connection and updating can emerge rather than be ineffectively coerced.

Occasionally, we may have the opportunity to introduce new information that challenges someone's existing beliefs in a way that might benefit them. However, how we do this matters greatly. Thoughtful belief updating means offering new perspectives without demand by presenting them as an invitation rather than an imposition. It requires respecting a person's autonomy in how they process and integrate new information and consider effective timing, introducing new ideas when the person is open and ready, rather than forcing them in a moment of distress or resistance.

A couple of guiding principles for navigating this area include practicing restraint in offering advice, framing insights as questions rather than directives, and sharing how your system perceives the situation while allowing space for the other person to interpret and respond in their own way, fostering mutual understanding rather than control.

## The Power of Repair: How Safety Grows Through Imperfection

**Fortunately, safe containers actually benefit from a lack of perfection.** When small breaks in trust occur, they provide an opportunity to repair those ruptures. Addressing these moments in a timely and thoughtful way strengthens the relationship, reinforcing the understanding that it can withstand challenges like misunderstandings or inadvertent harm. Maintaining the connection depends on limiting the impact of ruptures and repairing them quickly to support lasting stability and resilience.

## Building Internal Stability in Unstable Times

**Even when the world around us feels unpredictable or unsupportive, we still have the opportunity to cultivate a sense of safety from within.** From the perspective of active inference, we do this by reducing free energy (the gap between what we expect and what we experience) through intentional actions that increase coherence and predictability. A safe container, in this sense, works by offering consistent and trustworthy cues that help our internal models stabilize. When that external predictability is missing, we can instead introduce small, internal sources of structure and rhythm to help settle our system and reduce free energy. Psychology supports this by showing how small, intentional actions can serve as regulatory anchors. This may include engaging in daily rituals, grounding through the body, naming emotions, carrying a tactile object, playing familiar calming music, or focusing on a single aspect of life that feels steady. These practices offer reliable inputs to the brain, helping it orient and stabilize in the absence of broader external predictability. Small, consistent actions like feeling the ground beneath your feet, writing an intention each morning, or speaking compassionately to yourself create signals of predictability that strengthen the sense of safety in your generative model. These actions help reduce uncertainty internally, making it possible to continue updating beliefs and behavior with less overwhelm. In this way, even amidst outer chaos, we can create micro-containers of safety within our perception-action cycles. This offers us greater stability for our systems to stay engaged, update, and respond with greater clarity and resilience.

## Attachment Theory Through the Lens of Active Inference

Attachment theory, deeply influenced by John Bowlby and extensively articulated by Jeremy Holmes, highlights the profound significance of early relationships in shaping our lifelong emotional strategies. Integrating Holmes' perspective with active inference provides a nuanced understanding of attachment as a process deeply embedded in the prediction, management, and reduction of uncertainty.

From an active inference perspective, secure attachment can be viewed as the establishment of a reliable "safe container." Holmes emphasizes that when caregivers sensitively respond to a child's emotional cues, they foster internal predictive models that reliably anticipate safety and support. This consistent responsiveness reduces the child's internal uncertainty—known in active inference as free energy—and thus creates a stable psychological and physiological environment.

Research on animals, such as Harlow's classic experiments with monkeys, vividly illustrates the necessity of comforting, predictable presence for emotional regulation. Monkeys preferred the comfort of a soft, cloth surrogate mother even when nourishment was available only from a wire surrogate, emphasizing the critical role comfort plays in creating reliable predictions of safety.

Holmes extends this concept into human development. Securely attached individuals maintain an internal working model that predicts positive, reliable responses from significant others, effectively minimizing uncertainty and associated emotional distress. In contrast, insecure attachment, expressed as either avoidant or anxious/ambivalent, reflects maladaptive predictions about relational security. Avoidant individuals tend to expect emotional unavailability and consequently distance themselves to minimize expected disappointment. Anxious/ambivalent individuals, conversely, anticipate inconsistency, causing persistent anxiety and excessive vigilance as they attempt to achieve reassurance.

Fortunately, as humans who are generative models, we have the ability to shift our attachment styles. We can intentionally reshape our internal predictive models and how we anticipate emotional experiences in relationships. Some ways to do so include:

**Create a Safe Container:**

Engage in relationships, therapy, or supportive environments that offer consistent empathy, reliability, and emotional availability. Experiencing these consistently can recalibrate expectations and reduce uncertainty.

**Increase Reflective Awareness:**

Foster reflective functioning or mentalization—the ability to notice and understand your own emotions and the emotional states of others. Practicing mindfulness, journaling, or self-reflection helps you consciously adjust outdated or maladaptive predictive models.

**Update Predictive Models Gradually:**

Through repeated experiences of safety, responsiveness, and trustworthiness in relationships, you build evidence that challenges previous expectations of rejection, abandonment, or inconsistency. Active inference suggests that this steady exposure helps internalize new, healthier relational predictions.

**Therapeutic Relationships as a Tool:**

Holmes emphasizes the therapeutic alliance as a prime example of reshaping attachment. Engaging with a therapist who consistently provides emotional safety helps you internalize a new, secure model of relating.

## SPARK

Think about the relationships you're already engaged in and the ideas from this chapter. Did anything stand out to you? If so, consider revisiting that section for deeper reflection. Journaling or taking a reflective walk can also help integrate what you've learned. Given this new knowledge and understanding, what actions feel meaningful to explore or integrate more deeply?

# FREE ENERGY IN THE SYSTEM: WHY OTHERING HURTS US ALL

# FREE ENERGY IN THE SYSTEM: WHY OTHERING HURTS US ALL

One of the goals of this book has been to focus on what active inference and the free energy principle can help illuminate about our shared human experience and insights that often transcend background or belief, and invite a sense of coherence we can feel together. While conversations about divergent applications of this science are important, many of those discussions feel especially charged right now—so much so that they can make effective conversation more difficult. It's just as important to find moments to come together. Otherwise, we risk fracturing so far that the bridges between us become too wide to sustain. They collapse under the weight of unpredictability, uncertainty, and unsustainable free energy loads. This chapter is written with that intention in mind. Additional conversations about the specifics are important, and while they don't take place in this chapter, it's our hope that this chapter can contribute to the larger ongoing discussion. This chapter also isn't written from the lens of a policy or social justice expertise, but rather from a place of personal experience and active inference grounding. It offers a glimpse into how "othering"—the deep, often unconscious division between "us" and "them"—ultimately destabilizes everyone's active inference cycles, limiting our ability to adapt and unfold.

**ANNOTATION**

Ultimately, this benefits our individual and collective active inference cycles.

## Personal Story: Portugal

I'm American-Portuguese. My father immigrated to the U.S. in his 20s, and I was raised in the States, but with regular visits to our small village in northeastern Portugal. Growing up, I didn't frame those experiences in terms of "us" and "them." Life was simply different. I experienced something deeply relational and nourishing—playing soccer in the square while elders talked nearby, walking animals to pasture, harvesting food from the land. It was simple. It was connected. It met a set of needs I was out of touch with in the States.

But I don't believe any single group of people has it all figured out. In Portugal, gender imbalances can be palpable. As a girl, I was told not to mention my brother doing dishes because other boys would mock him. As a woman, I've often encountered dishonesty from men (particularly around intentions for commitment and boundaries) in ways that have been marked by duplicity and felt disorienting. It's a system that burdens everyone it touches, men included, under the weight of outdated roles and norms.

And yet, one of the more unexpected observations came from social settings in Portugal as an adult. When I introduce myself as someone from San Francisco, I'm often welcomed with curiosity and perceived value. When I mention I'm American-Portuguese and that my father is from a northern village, the tone shifts. Sometimes subtly. Sometimes not. It's a form of "epistemic mistrust" that Peter Fonagy writes about: the immediate dismissal of information as irrelevant or untrustworthy, simply because of the source. In this case, my perceived value and the weight of my opinions sometimes appear disvalued when my parental lineage is linked to a small northern Portuguese village.

## Applied to Active Inference and Free Energy

Let's ground this in physics and science. When I feel I must hide or edit who I am to be accepted in a social setting, my system is working harder. This is free energy in action. My generative model must account for conflicting predictions: "I belong here" versus "I may be rejected." That conflict drives surprise. It increases uncertainty. And to keep up the appearance of fitting in, I must expend more cognitive and emotional energy.

But it's not just about me. The people who dismiss or miscategorize me are also operating under overly rigid beliefs, often making it more difficult for them to adapt. Their belief that someone from a village (or anywhere unlike their own background) is less worthy of attention is a kind of frozen narrative. That rigidity prevents updating, limits exploration, and thus, limits adaptability. Ultimately, placing limits on their own active inference cycles and disrupting their ability to experience ease and adaptability in an ever-changing world. These rigid priors are not irrational, but adaptive strategies that once helped the system cope with uncertainty, even if they now limit flexibility.

A third layer reveals itself in family dynamics. Some of my extended family is from the village, some from the city. At times, I've needed to set boundaries because these interactions became energetically costly. But the fallout from those boundaries shocked me. Only later did I realize I'd inadvertently triggered a deeper system: one where my stepping back wasn't interpreted as a boundary for health, but as a rejection of the individual's identity and perceived worth. Again, we see how when we're unable to integrate new contexts and understandings we

**ANNOTATION**

Drawing from Fonagy's work, this might be seen as "epistemic petrification": a rigid, unduly precise, internal model that can't integrate new, disconfirming data.

**Navigating Uncertainty**

"When I feel I must hide or edit who I am to be accepted in a social setting, my system is working harder.

can create loops of misunderstanding and distress that limits, to varying degrees, all involved.

This pattern repeats elsewhere. In conversation with a young Portuguese graduate of a top U.S. university I sensed a struggle in conversation around gender equality. His inability to update, I sensed, wasn't rooted in ideology alone. Having spent time in the US, I couldn't help but wonder if part of it stemmed from unintegrated experiences he had in the US. Perhaps at University he felt subtly "less than" in certain conversations because he wasn't American. Over time, these experiences may have taught his system to guard against unfamiliar ideas, not out of stubbornness, but as a subtle form of protection. As Fonagy and colleagues suggest, when people feel unseen or misrecognized, their systems learn to protect themselves by refusing incompatible inputs. Not because they're irrational, but because it once kept them safe. In other words, these loops of misunderstanding and distress don't just stay contained—they echo across people and places, adding tension and free energy to the broader system we all share.

## What the Science Suggests

Karl Friston reminds us that "we're already near Bayes-optimal," but most of our inferences run subpersonally. In other words, our brain is like an incredibly powerful and accurate machine, but it can run automatically, using outdated software. To update, we need to become aware of our interpretation processes and regain adaptability. This includes recognizing when our beliefs about ourselves, others, or the world are too firm. When they no longer fit the data, but we cling to them anyway, we get stuck. Like being caught in the same mental loop, even when the world around you has changed.

The way out? Create space for updating within our own and collective active inference cycles through trust. Specifically, Fonagy's work highlights the importance of epistemic trust. That our and others' capacity to take in new information greatly increases when it comes from someone who sees us, respects us, and aligns (at least momentarily) with our internal model. Without that trust, we tend to block out new data before it can change our minds or update our beliefs. With it, our system softens—and meaningful updating becomes possible.

And so, while I don't have answers for all forms of injustice, I know two things from science: First, understanding our own inference systems can rapidly shift our ability to update. And second, that we and others do this much more effectively when we feel safe.

So maybe this chapter is simply a gesture of safety. A reminder that conversations like this need to be had with empathy in order to be effective. That othering doesn't just hurt those cast as "other." It increases free energy across the entire system. It limits us all.

## SPARK

- ✦ Think back to a time when you felt like an outsider—whether because of your identity, beliefs, background, or simply how you showed up. How did your system respond? What predictions did it begin to make about future social situations?

- ✦ How can you use this understanding to shift others' experiences of othering?

# NEURO-DIVERSITY: A VALUABLE HUMAN CONSTRUCT

# NEURODIVERSITY:
# A VALUABLE HUMAN CONSTRUCT

Science doesn't just explain what is. It also invites us to explore what could be. The research and writing of this book enabled me to collect especially inspiring facts and insights that provide opportunities to shift our perspective on human potential, our place in the world, and reduce our free energy. One such fact stands out: only about 1.5% of the human genome codes for proteins—the essential molecules that form our bodies and make our cells function. The other 98.5% includes unexpressed potential. This means that within us exists a vast library of possibilities, capacities that remain dormant.

We can see hints of this potential in extraordinary individuals such as child prodigies who master complex musical compositions before they can read, young mathematicians who solve problems that stump seasoned professionals, and doctors who, before reaching adulthood, perform life-saving surgeries. Think of Wolfgang Amadeus Mozart composing symphonies as a child, Akrit Jaswal performing surgery at age seven, or Terence Tao, who scored a perfect SAT math score at eight years old. These individuals represent flashes of what potentially lies encoded within our species. And while such remarkable abilities may be rare, their very existence is hopeful and inspiring. They remind us that humanity's potential is vast, filled with possibilities that we may not yet fully understand, but that offer hope and inspiration for what is to come.

Neurodiversity is another fundamental characteristic of humanity. Current estimates suggest that approximately 15-20% of the population is neurodivergent in some way. This includes conditions such as autism, ADHD, dyslexia, synesthesia, and other variations in cognitive and sensory processing. When we take this into account, it becomes clear that either we ourselves or someone we are close to likely experiences the world differently from a neurocognitive perspective. That difference extends beyond mere preference or personality—it means that the way their active inference cycles navigate the world, how they sense, perceive, and infer meaning is unique, private to each of us, and necessarily diverse.

Personally, my experience of autism is a key ingredient of this book. My "wellness journey," which involves learning how to respond to previously overwhelming flows of sensation, has required me to develop unique coping strategies, such as understanding active inference and free energy. Because of this experience, I have the opportunity to share these insights with you, helping you make use of your own experience—whether it feels like a gentle stream or a roaring waterfall. However, my experience, and that of many autists, is fundamentally different from that of neurotypical individuals.

To create opportunities for society to benefit from cognitive diversity, we need to ask others to approach our experiences with empathy and understanding. For example, environments filled with unpredictable stimuli such as loud noises and fluctuating social cues require more processing effort, seem different, or respond in ways that may be perceived as unexpected in group settings. This can be especially challenging in my case as someone in the "wellness" space, people naturally assess my credibility based on my behavior. However, for this bridge to function, both with myself and other autists, is for people to recognize that my behavior may be different or sometimes awkward—not because we lack ability, but because our systems operate differently, constantly managing an intense sensory and cognitive load. Compassion and understanding go a long way in fostering an environment where autists can share ideas and insights, ultimately contributing their unique perspectives to benefit us all. In the meantime, others, like myself, continue to refine our coping strategies, seeking authentic ways to bridge the gap between my experiences and others. When we all stretch to build a bridge across the gap, we create greater collective opportunities to benefit from a network of bridges.

A quick note before we jump further into this discussion. Diagnoses are best understood as models rather than rigid categories. They offer a framework for describing patterns of cognition and perception, but do not encapsulate a person's full experience or potential. I like to think of humans as a dynamic recipe rather than a fixed dish. Cognitive models are just one ingredient, shaped by countless other factors like environment, experience, and adaptability. The way these elements interact determines the outcome, much like how a well-balanced ecosystem thrives not just because of its diversity but through the functional integration of its elements. We include these models here in that light, as tools for understanding differences in sensory processing, attention, and social interaction—not as rigid structures, but as adaptable paradigms that help us foster understanding and innovation within society.

**Navigating Uncertainty**

## Neurodiversity Through the Lens of Active Inference

The free energy principle and active inference cycle continue to act as anchors for us to weave deeper understandings, this time acting as a lens for understanding neurodiversity. As we've discussed, cognition is not about passively receiving the world as it is but about actively generating predictions, testing them, and updating our understanding (the arrows point both ways). From this perspective, neurodiverse cognitive processes are not 'errors' or 'deficits' but adaptive strategies—ways of perceiving and interacting with the world that expand humanity's ability to respond to different environments and internal states.

Neurodivergence, in this light, can be seen as a form of 'computational diversity'. Diverse ecosystems flourish when species play distinct roles, like bees pollinating and trees producing oxygen. In the same way, computational diversity builds resilience by combining varied models or cognitive strategies that work together effectively. This requires functional integration; diversity alone would not ensure adaptability or stability, much like a sports team where players can't communicate or a jazz band where musicians play independently instead of harmonizing. However, when we create functional communication and integration strategies, this interplay brings increased capacity and adaptability, just as a well-coordinated sports team performs better than one where players don't communicate, or a jazz band creates richer music when musicians harmonize, riffing off each other, rather than playing independently.

Neurodivergence manifests in unique ways for different individuals, shaping how they perceive, process, and respond to the world around them. Some

---

**ANNOTATION**

A metaphor for the different ways humans minimize uncertainty, process sensory input, and engage with the world.

individuals' active inference cycles may place heightened emphasis on sensory precision, leading to deep focus in certain areas but sensory overload in others. Others may rely on broader patterns and rapid inference, leading to quick adaptability but sometimes missing finer details. These variations enhance our collective ability to navigate and shape our environments, at various levels including family and work.

For example, a parent or teacher with heightened sensory precision may be especially attuned to a child's nonverbal cues, such as subtle shifts in mood, body language, or sensory sensitivities. This heightened awareness by the caregiver enhances nurturing environments for developing children, fostering positive outcomes like increased problem-solving approaches and increased emotional intelligence and resilience.

Similarly, in the workplace, diverse cognitive styles drive innovation, problem-solving, and creativity. Neurodivergent individuals often bring heightened attention to detail, increased memory retention, and unique problem-solving approaches. For example, dyslexic thinkers frequently excel in big-picture analysis, while individuals with ADHD may thrive in dynamic, fast-paced environments, bringing fresh perspectives and adaptability to problem-solving. More broadly, a 2017 Harvard Business Review report noted that companies integrating neurodiversity can benefit from greater innovation, efficiency, and adaptability.

Our active inference cycles are shaped by the signals we receive and the predictions we make about the world. When we create space in a balanced and functional way to engage with neurodivergent individuals, we are exposed to different modes of sensing, perceiving, and predicting, which refine our internal models. This process strengthens both individuals and the system as a whole, much like biodiversity stabilizes an ecosystem. Through integrating collective adaptability, we build a society that is more resilient, innovative, and capable of handling complexity with greater ease.

## Some Types of Neurodivergency

### Autism: A Different Mode of Active Inference

Through the lens of active inference, autism can be understood as an alternative way of processing information, shaped by differences in sensory precision and predictive modeling. Specifically, autism has been framed as a matter of precision weighting, incorporating a different balance between sensory evidence and prior expectations. Individuals on the spectrum often experience heightened sensory precision, meaning their brains assign greater weight to raw sensory input rather than filtering it through prior knowledge and context. This can lead to a fundamentally different experience of the world, where sensory input is more vivid, and details are more pronounced. For example, when making the choice to drink from a fountain versus a water hose, technically it's possible to do so for both, but the strategies for doing so and potential for getting overwhelmed with water are very different. Ultimately, this often requires autists to develop an increased capacity to respond effectively, yet allows for extraordinary attention to detail and heightened pattern recognition.

Beyond autism, several other forms of neurodivergence provide unique opportunities for understanding valuable cognition approaches through active inference and free energy minimization. While we are unable to include all of them here, conditions such as ADHD, dyslexia, and synesthesia are some examples that each offer distinct ways of interacting with the world.

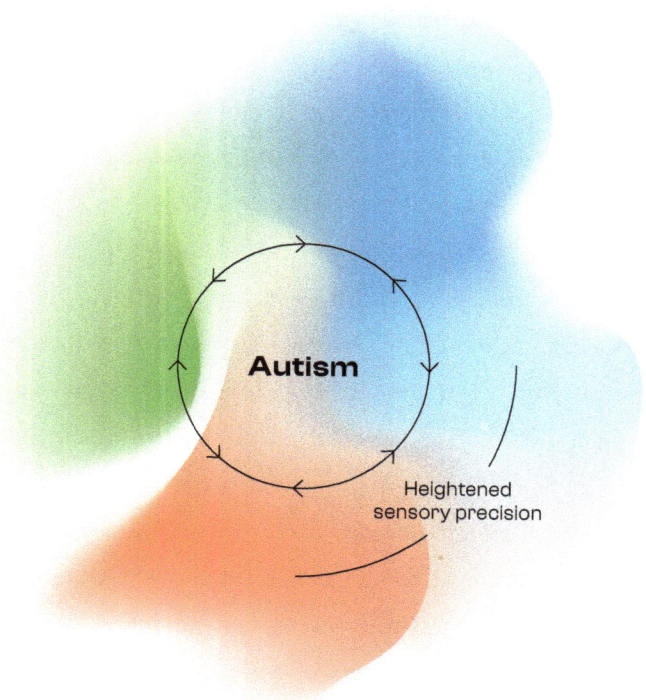

**Autism**

Heightened sensory precision

### ADHD: A Different Dance of Attention and Energy

One way ADHD can be viewed is as a system optimized for rapid adaptation rather than sustained focus. Individuals with ADHD often demonstrate a heightened responsiveness to novelty, leading to quick shifts in attention and high energy levels. In active inference terms, their predictive models may prioritize fast-paced learning and exploration over rigid structure, making them highly adaptable in unpredictable environments.

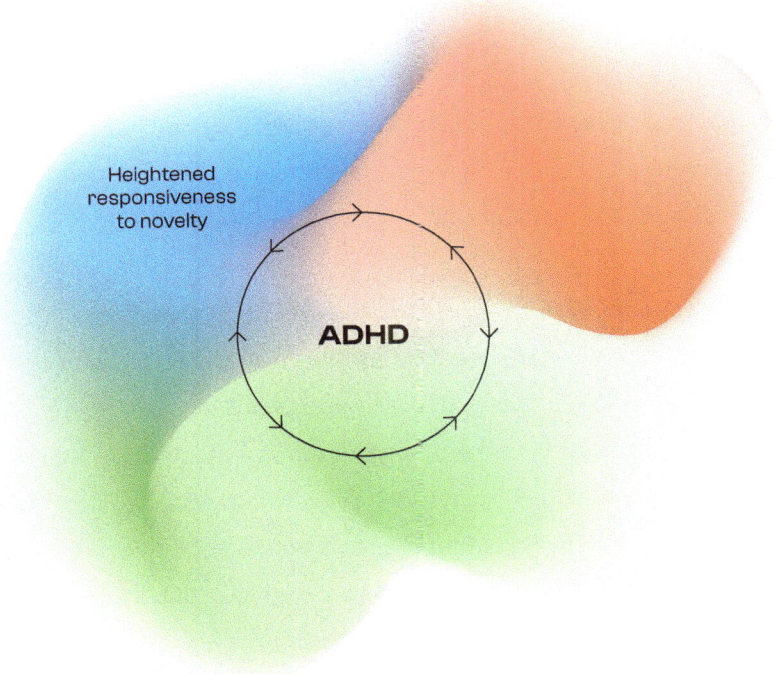

This can present challenges in traditional educational and workplace settings, often requiring system flexibility and development of coping mechanisms such as structured routines. It also offers beautiful opportunities for increased creativity, spontaneous problem-solving, and innovation. The ability to detect patterns in chaotic environments and hyper-focus on subjects of interest can lead to groundbreaking ideas and high levels of productivity. Implementing strategies such as breaking tasks into smaller steps, using external reminders, incorporating movement into daily routines, and fostering interest-based learning enables many individuals with ADHD to enhance focus and productivity while working with their natural strengths. In addition to fostering empathy, we can integrate these valuable perspectives and experiences into our collective knowledge by supporting the need for increased movement, clear communication, and flexible approaches to productivity for some.

### Dyslexia: A Different Lens of Detail vs Holistic Thinking

Dyslexia is often associated with difficulties in reading and language processing, but it also often comes with cognitive advantages, particularly in big-picture thinking and pattern recognition. Individuals with dyslexia tend to process information in a more holistic and associative manner rather than through linear sequencing. From an active inference perspective, their prediction models may favor global connections over fine-grained textual details, allowing them to excel in fields like engineering, design, and storytelling. While traditional literacy models can present challenges, shifting educational frameworks to embrace multimodal learning can empower dyslexic individuals to utilize their strengths, providing gifts back into society. Implementing strategies such as using audiobooks, employing speech-to-text tools, focusing on visual learning techniques, and breaking information into structured segments, many individuals with dyslexia can enhance their comprehension and learning experience while leveraging their natural strengths. Others can support people with dyslexia by offering flexible communication options, minimizing unnecessary written demands, and recognizing the value of big-picture thinking.

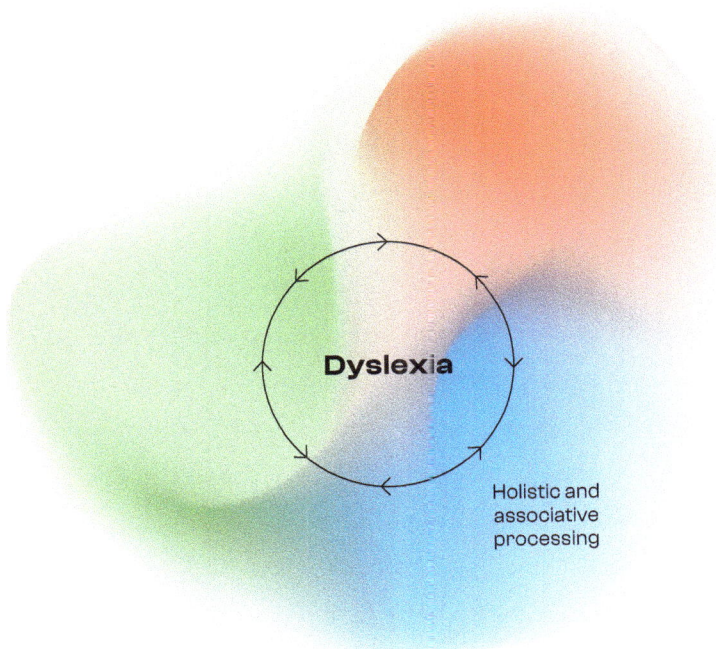

### Synesthesia: Blended Senses and Enhanced Experiences

Synesthesia is a neurological trait in which sensory inputs are automatically linked, such as seeing colors when hearing music or associating numbers with specific textures. This unique way of processing information suggests a brain that may integrate sensory signals more extensively than in neurotypical individuals. In active inference terms, some synesthetes experience a world where predictive models fuse multiple sensory modalities, potentially reducing uncertainty by reinforcing associations between disparate stimuli. This heightened connectivity can support enhanced memory, creativity, and artistic expression, and introduce novel ways of experiencing and interpreting information.

For example, a mental health professional with synesthesia could experience emotions or spoken words as colors, textures, or even musical tones, allowing them to perceive subtle emotional shifts in their clients more intuitively. This could, in some cases, help them recognize underlying emotions that might not be explicitly stated, providing deeper insight into a client's mental state and enhancing therapeutic techniques. Some people with synesthesia channel their abilities through structured associations or creative expression, while others find mindfulness helpful in managing sensory overload. These approaches can enrich both personal and professional experiences. For others, support often comes through openness to their perspectives and encouragement of creative outlets, rather than holding on to rigid ideas about how things "should" be perceived.

While these examples highlight very real strengths across ADHD, dyslexia, synesthesia, and other forms of neurodivergence, those strengths also coexist with challenges. Supportive environments make it possible for individuals to thrive. They provide space to work with differences in ways that lower unnecessary barriers and let unique contributions create value.

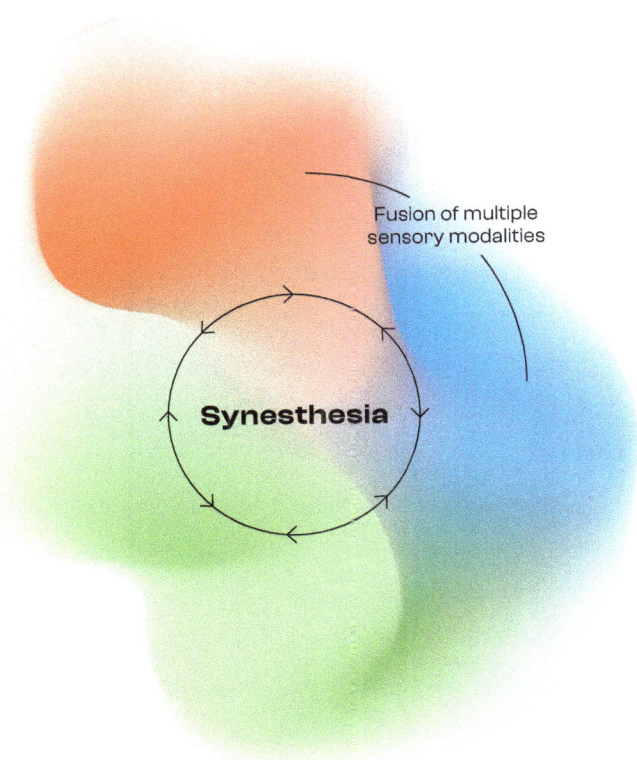

Fusion of multiple sensory modalities

Synesthesia

## A Personal Perspective

Science provides an essential foundation for understanding neurodiversity while personal stories provide the opportunity to enhance emotional understanding. We often remember and relate to stories more than data points. That's why I choose to share my personal perspective here, not because it is the universal experience of neurodivergence, but because it offers one way of seeing these ideas come to life.

I have had the privilege of knowing and working with neurodivergent individuals, both personally and professionally. For me, this has often felt familiar and even resonant; in part because, as we discussed, I'm likely neurodivergent myself. Connecting with others who share similar experiences often feels natural, effortless, and rewarding. However, I have observed that for some, engaging with neurodivergent individuals can be more challenging—not due to any fault of their own, but because their active inference cycles interpret my way of interacting as unfamiliar.

This is further complicated by the societal tendency to associate autism and neurodivergence predominantly with males. As a female, my neurodivergent traits are not only less recognized but also contrast with the often unstated expectation that I exude traditionally feminine social energy. The interaction patterns many are accustomed to may not align with neurodivergent ways of perceiving and predicting the world. Recognizing this has been incredibly helpful, allowing me to be more intentional about communicating my experiences and behaviors to others. By openly sharing how I process information, I create opportunities for mutual understanding and reduce potential misinterpretations.

For example, when I miss a social cue, it is usually due to the overwhelming amount of information I am processing rather than an intentional slight. Over time, I have refined my ability to bridge these differences in ways that foster smoother interactions without excusing all socially challenging behaviors. I have been pleasantly surprised and incredibly grateful for the positive responses I have received and how effectively sharing my experiences has facilitated a deeper connection and ease for many involved.

## The Question of 'But Are You Really?'

At times, people have asked me directly, "But are you really on the spectrum?" This question can sometimes require conscious effort not to feel frustrated. Since you're reading this book, you likely recognize that our ability to understand another's active inference cycle is inherently limited; we are best equipped to interpret our own experiences while approaching others with curiosity and non-judgment. Thus, I choose to see this question not as skepticism but as a genuine attempt to understand—a moment of curiosity that, when met with openness, can foster meaningful dialogue and connection.

**DEFINITION**

This is what cognitive scientists call the theory of mind: our capacity to infer that others have beliefs, intentions, and experiences that may differ from our own.

From a scientific point of view, "But are you really?" is an intriguing question. How would you know, and how would I? This touches on a fascinating and complex issue in philosophy of mind. If my way of making sense of the world (my active inference cycle) is unique to me, then so is my model of you. I can't access your private thoughts, your experiences, or your intentions. But if you behave in ways that are familiar to me, I can make an educated guess that what it's like to be you isn't so different from what it's like to be me.

However, this process isn't always straightforward, especially when considering neurodiversity. For example, a neurotypical person might interpret a lack of eye contact as disinterest or rudeness, while an autistic individual might avoid eye contact to manage sensory overload. If neither party recognizes this difference, misunderstanding can arise. But through engagement and curiosity, we can gather more understanding about each other's experiences, reducing uncertainty and fostering mutual understanding. When it comes to my own experience, I could approach this question from several different angles. My system is measurably different in ways that resonate with neurodivergence. In a lab setting, I distinguish color differences invisible to 90% of the population. My auditory perception is similarly attuned; I used to help the second sopranos find challenging notes in high school choir. In crowded events, such as soccer games, I often react to events before the crowd. If a goal is imminent, I sense it earlier; if a shot is doomed to miss, I don't cycle through the false hope others around me do. This early evidence of my different perception led me to adjust—initially, by suppressing my natural reactions to fit in. Over time, I've become more comfortable letting my responses be authentic, especially when they fall within the larger socially acceptable range of a soccer game.

These differences continued into my professional life. I initially struggled with understanding the value of safe containers in group dynamics. Over time, I learned (partly through kind and patient neurotypical mentors) that my processing speed was different than others. Changing my pace in conversations and meetings helped create a shared space where ideas could be meaningfully exchanged.

This realization also deepened my empathy. **If I ask others to accommodate my neurodivergence, it requires I also recognize and respect the cognitive styles they bring.** I have come to genuinely value the strengths neurotypical individuals often bring, such as strong social intuition, the ability to synthesize broad contextual cues, and a natural ease in group dynamics, all of which help create more effective interactions and collaboration. While I have not sought a formal diagnosis, the evidence strongly suggests that I experience the world in ways consistent with autism, and recognizing this has provided me with valuable insights into how I navigate life.

I also acknowledge that my experience is shaped by a mix of genetic, epigenetic, and environmental factors. It's not entirely clear to me, nor do I believe it needs to be, which elements most influence my active inference cycle. Rather, what matters is understanding that my mind operates differently, valuing the unique strengths that both neurodivergent and neurotypical individuals bring, and learning to communicate my differences in ways that foster understanding and reduce free energy in my interactions. At times, I have wished for a more neurotypical experience, but I genuinely don't believe one way of being is inherently better than the other. Each comes with its own challenges and gifts. It has been, and continues to be, a journey of acceptance and self-awareness.

## Some Additional Reflections

Diversity in active inference cycles is essential for a thriving society. When detail-oriented thinkers and big-picture strategists possess the skills to collaborate effectively, efficiency and innovation flourish. In families and communities, neurodivergent individuals can provide unique perspectives that enhance problem-solving, deepen emotional intelligence, and create richer, more adaptive relationships.

Rather than striving for uniformity, creating space for cognitive diversity allows us to create environments that support different processing styles. This means ensuring that children today have the space and tools to develop according to their unique strengths, rather than being forced into conventional molds that may not serve them. Viewing this as a long-term investment fosters a future where innovation flourishes, communities become more adaptive, and societies grow more resilient.

Ultimately, neurodiversity is a powerful source of hope—a reminder that humanity's potential is vast, its strengths are varied, and its ability to adapt and thrive is enhanced by the very differences that set us apart. Recognizing and nurturing these differences enriches individual lives while contributing to a world that grows more innovative, intelligent, and resilient.

# SPARK

- Whether or not you identify as neurodivergent, take a moment to reflect on how your mind works.

- Is there a way you think, feel, or process that has often felt different—or been misunderstood by others?

- What strength might be embedded within that difference, even if it hasn't always been recognized?

- How might you begin to cultivate your life in a way that supports and amplifies this strength rather than suppressing it?

- Beyond your own experience, how might you help create space for others to do the same while honoring their unique methods and ways of navigating the world?

# NAVIGATING RELIGION, SPIRITUALITY, & PITFALLS

# NAVIGATING RELIGION, SPIRITUALITY, & PITFALLS

Just as our cognitive styles shape how we sense and respond to the world, our spiritual and existential models also guide how we generate meaning and navigate uncertainty. This book seeks to foster dialogue and shared understanding within active inference cycles, easing free energy and strain both individually and collectively. The robust science that supports these concepts is extensive. For this reason, we haven't focused on areas where some may perceive tension with their religious beliefs, such as evolution. Perhaps you haven't even noticed that this hasn't been included.

There is a point where science ends and our human experience takes over, either because we haven't gotten there yet or because it lies beyond an inaccessible horizon. Where we draw that line within our own cycles and understandings is deeply personal. As we've continued to see, it only makes sense that different cycles arrive at different conclusions shaped by factors like when and where we are born. This chapter explores where active inference allows for diverse perspectives, how we can benefit from these perspectives as humans, and how to avoid ineffective inference loops.

## Opportunity and Boundaries for Religion in Society

I am incredibly grateful for the opportunity to have connected with people from many religious and spiritual paths while researching this book. This diversity of thought includes perspectives such as "I'm a tree hugger and believe God will return one day," "I believe in Creationism," "I use horoscopes," and "Science implies our lives are meaningless." Given the broad spectrum of beliefs and research findings, two key principles emerge to help us come together:

- Religion should not be used to escalate conflict.
- Religious beliefs should not be used to discredit robust science. Simultaneously, those who align with science can make space for personal beliefs in others that don't fully align with scientific consensus.

The first point is likely self-explanatory. The second point, however, may benefit from a bit more exploration. For example, creating a policy that assumes God will rescue us from climate change would be problematic according to these two principles because it dismisses scientific consensus and impacts our collective future. However, believing in Creationism and a future redemption is a personal choice that does not necessarily interfere with general scientific inquiry or the two possible guiding principles above.

Both scientific and non-empirical beliefs play a crucial role in shaping how we understand and engage with the world. These perspectives (whether grounded in faith, philosophy, or empirical research) help us build shared frameworks for navigating life. Exchanging and integrating these viewpoints builds a society that is more stable and sustainable. Rather than being at odds, science and non-scientific perspectives can contribute in different but often complementary ways to the structures that allow society to function cohesively.

## Benefits of Religion and Spirituality

Whether you're part of a formalized religious group like Catholicism, Judaism, Hinduism, Buddhism, or Islam, identify as a generally spiritual person, or only believe in what can be demonstrated through science, evidence points to many religious and spiritual practices offering potential benefits for humans. For some, these benefits might be explicitly stated within their Faith. For others, we can look through the lens of evolutionary science, which supports the idea that humans have traditionally evolved with religious experiences as a key cornerstone of our existence. An example of religion and spirituality with well-documented benefits is meditation in Buddhism, which has been shown to reduce stress and improve cognitive function.

While we haven't yet explored practices like prayer or chanting in Christianity and Islam, research suggests they are linked to enhanced emotional resilience and stronger social bonds. From an active inference perspective, these practices provide predictable, rhythmic sensory input that helps reduce internal uncertainty and regulate emotional and interoceptive states. When performed communally, they also enhance social synchrony, increase collective coherence, and reduce uncertainty in interpersonal interactions.

In another similar, but distinct example, we can note that the Jewish practice of Shabbat also aligns with principles we discussed in Chapter 18, where we explored how intentional rest can restore physiological and cognitive balance, reduce internal uncertainty, and support more accurate inferences. Such structured downtime can provide an opportunity to recalibrate internal models, improve emotional regulation, and enhance overall system coherence.

Yoga in Hinduism is yet another example—studies have shown it improves both physical and mental well-being and offers structured, embodied practices that integrate breath, movement, and focused attention. This can reduce prediction error by aligning internal bodily states with external actions.

One final example is indigenous spiritual rituals that reinforce cultural identity and emotional healing. From a scientific perspective, such practices have been linked to reduced stress, increased resilience, and stronger community ties. These outcomes benefit our active inference cycles by supporting coherent belief updating and emotional regulation. However, **we potentially benefit most when such practices align with our understanding of active inference and the broader scientific context**: otherwise, they may interfere with adaptive learning and reduce our ability to flexibly update internal models in response to new information.

In short, scientific evidence supports many religious and spiritual practices in reducing uncertainty, strengthening resilience, and supporting overall well-being—illustrating the profound role spirituality can play in enhancing human health and social cohesion. Whether we approach it from the lens of science or religion, both perspectives highlight potential value in religion and spirituality. Being human is incredibly complex, and finding purpose and meaning

"There is a point where science ends and our human experience takes over, either because we haven't gotten there yet or because it lies beyond an inaccessible horizon.

can be challenging—especially when evolutionary science describes suffering as a natural mechanism for adaptation and survival but does not assign it intrinsic meaning. Ultimately, where we land on this spectrum is deeply personal, but acknowledging and refining our beliefs with awareness can lead to greater coherence and a more meaningful existence.

As a scientist who was once disillusioned by a specific religious community in my childhood, I spent much of my adult life unwilling to welcome any form of spirituality. However, over time, I have come to hold onto a spiritual belief, lightly focusing on elements that align with active inference and the free energy principle. For me, this was not something I actively sought out; rather, it felt like an acknowledgment and necessity of establishing a functional set of beliefs within today's society. Additionally, I make time for spiritual experiences and deeply appreciate when others share theirs with me. As a scientist, I can't tell you if it's because these practices hold some sort of fundamental truth that I'm connecting to, or because I'm acknowledging what it means to be human from an evolutionary and experiential point of view. Whatever the mechanism, creating space for it has fostered significantly greater resilience, meaning, and purpose in my life.

## Horoscopes

Horoscopes can serve as tools for reflection, purpose, and meaning. From an active inference perspective, we recognize the importance of continually refining our understandings. Scientifically speaking, it is conceivable that cosmic energy has some impact on organic matter, including humans, that we don't yet fully understand. As Carl Sagan famously said, "We are made of star-stuff." Scientists have measured fascinating and unusual cosmic phenomena like gravitational waves from colliding black holes. They've also captured neutrinos from exploding stars, mapped the invisible presence of dark matter through gravitational lensing, listened to radio pulses from spinning neutron stars (pulsars), and even imaged the shadow of a black hole using a network of telescopes spanning the globe. These feats involve detecting minuscule signals buried in cosmic noise, revealing a universe more dynamic and strange than we ever imagined.

However, horoscopes present a challenge: they are not rigorously studied nor are they updated based on actual outcomes. For example, astronomical reference points used in astrology have shifted. Earth's axis slowly wobbles over time in a cycle called axial precession, causing the zodiac constellations to drift roughly one sign every 2,000 years. This lack of refinement is concerning, especially when horoscopes are used for more than inspiration or self-reflection. Active inference emphasizes learning through prediction error minimization. Horoscopes typically don't update their predictions based on whether those predictions matched outcomes (i.e., there's no feedback loop), which means they don't refine their generative models in response to new information—a key feature of adaptive systems. Without adaptation to real-world feedback, using horoscopes is akin to relying on outdated medical practices for critical procedures.

Our cycles are naturally drawn to structures (like horoscopes) that appear to reduce uncertainty by offering explanatory narratives—tapping into our drive to minimize free energy. But when these narratives offer false certainty without

**Navigating Uncertainty**

adapting to real-world feedback, they can disrupt the very feedback loops that help us update our beliefs in line with reality. While this kind of quick uncertainty resolution may feel comforting in the short term, it can undermine long-term adaptability. Life is challenging, and while spiritual beliefs can offer comfort, if they remain static and unexamined, they risk trapping us in ineffective cycles—sometimes when we need clarity the most.

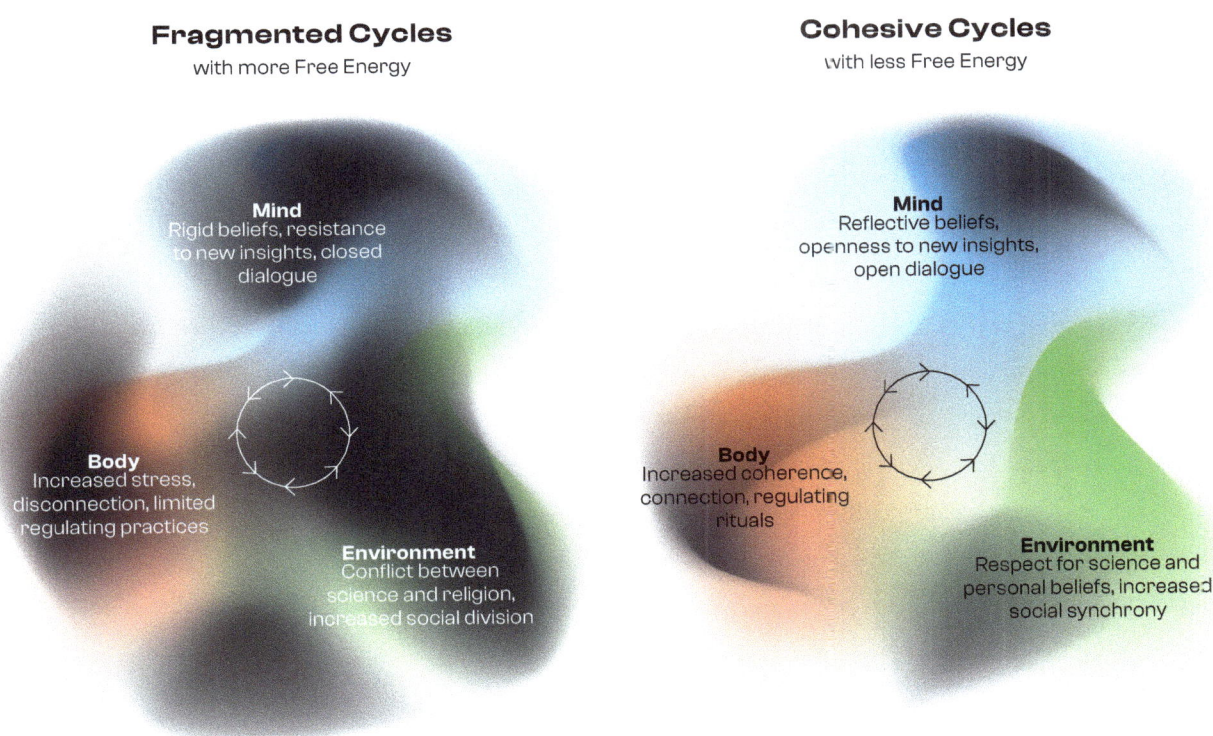

**Fragmented Cycles**
with more Free Energy

**Mind**
Rigid beliefs, resistance to new insights, closed dialogue

**Body**
Increased stress, disconnection, limited regulating practices

**Environment**
Conflict between science and religion, increased social division

**Cohesive Cycles**
with less Free Energy

**Mind**
Reflective beliefs, openness to new insights, open dialogue

**Body**
Increased coherence, connection, regulating rituals

**Environment**
Respect for science and personal beliefs, increased social synchrony

## A Possible Lens: Active Inference and Free Energy

As we saw above, many major religions incorporate elements that align with active inference and reduce free energy. It's also interesting to consider how our understandings of active inference and free energy can participate in the process of seeking intrinsic and experiential meaning and refuge within these practices.

Over time, religious interpretations shift to accommodate new understandings. Viewing these shifts through an active inference lens reveals them not only as historical changes but also as resources for shaping more adaptive, empowering practices today. For example, the Ten Commandments originally implied that people were property, yet we have long since abandoned the notion that it is acceptable to own another person. From an active inference perspective, human ownership decreased alignment between societal norms and opportunity for autonomy. Or the Hindu practice of Sati, where widows were expected to self-immolate, erasing their agency entirely, which has been outlawed. The caste-based restrictions in Hinduism, which also restricted role flexibility, and have been legally abolished. These shifts illustrate how belief systems, like individuals, can update their models over time, reducing uncertainty and increasing coherence within their communities. It's interesting to consider that, for those of us who choose to do so, we can refine the focus of our faiths today by emphasizing components that align with active inference and free energy, while preserving meaningful traditions and beliefs that foster a sense of connection, purpose, and resilience.

## Conversations Across Meanings

**Whether we embrace religious or spiritual beliefs or adhere strictly to scientific reasoning, our ability to function and benefit from society depends on bridging different perspectives.** This can be particularly challenging as these beliefs are deeply embedded within our active inference cycles, and shifting them can create significant ripple effects in our understanding and sense of stability. Often, this manifests as fear and discomfort, making such conversations easier to avoid than engage in. We previously explored the value of engaging in difficult discussions and the benefits they offer for both individual and collective active inference cycles, which you can refer to in Chapter 23. If we seek to cultivate greater meaning in our lives, making room for discomfort in these discussions can serve as a bridge to deeper connections and the enrichment of our perspectives, ultimately fostering a more adaptive and resilient worldview.

## SPARK

- Reflect on a ritual or routine (religious, spiritual, or secular) that brings you a sense of comfort or grounding.

- What does this practice offer your system? Decreased uncertainty, stillness, connection, renewal?

- Could you re-engage with this ritual in a more intentional way, or update it to reflect what you've been learning about your own self-knowing or active inference?

# TRAINING OUR BODIES FOR ACTIVE INFERENCE

# TRAINING OUR BODIES FOR ACTIVE INFERENCE

As you may be starting to notice, the seemingly simple, high-level process of active inference involves a vast number of levers and interaction points. In the previous chapter, we explored how our beliefs (whether grounded in religion, science, or spirituality) can either support or disrupt our systems. We can extend this idea to consider how our internal beliefs about ourselves, as well as our broader intentions, shape and influence how our systems function. When our intentions are shaped more by societal expectations than by the genuine needs of our own systems, we often find ourselves depleted, fragmented, and disconnected. This misalignment can quietly raise free energy, leaving us out of sync with what promotes coherence and vitality. This chapter continues that thread, but with a more bodily focus. Specifically, it invites a fresh perspective on movement and exercise. It also explores how these practices can enhance adaptive capacity and strengthen inner coherence through active inference.

Furthermore, we can remember our earlier discussions around cultivating safe containers within ourselves—spaces where we can hold our experience without judgment. This internal safety net allows for self-compassion during lapses or disruptions, and reinforces the understanding that it's the overall pattern of engagement that matters, not isolated moments of imperfection. Reframing physical practices through this lens of self-kindness and continuity can help ground them in deeper purpose and clarity, fostering a more adaptive alignment across mind, body, and environment.

## Adaptive Updating Through Embodiment

A key premise of this chapter is that bodily sensations shape how we experience, interpret, and respond to the world through our active inference cycles. These sensations act as core signals, offering insight into what's happening both within us and in the world around us, moment by moment. However, these signals are not always straightforward. For example, the sensation of tightness in the chest does not come with explicit cognitive labeling. Similarly, heightened arousal may correspond to either excitement or anxiety, depending on contextual cues and prior expectations. Moreover, some bodily sensations can be shaped by outdated or maladaptive predictive patterns. An old injury, for instance, may continue to produce bodily sensations that are misinterpreted as pain long after the tissue has healed, illustrating how past experiences can persist as entrenched priors within the system.

Movement and exercise provide a structured opportunity to engage with and update our explanations for these sensory patterns, reducing free energy from our systems. It allows us to notice, clarify, and potentially recalibrate bodily signals. One way to conceptualize this is to ask: What types of sensations do I want my system to be familiar with and fluent in interpreting? For example, I aim to cultivate a range of embodied experiences. These include grounded states that support embodied research, as well as more activated ones needed for public speaking, high-pressure environments, or even future biological processes like childbirth. These goals are highly individualized, reflecting the diversity of needs across human systems.

## A Model for Embodied Updating

To build on this, we can take inspiration from algorithm design and the active inference process itself. Just like algorithms update their models by cycling through new data, we can update our internal predictions by cycling through new experiences. Unlike algorithms, however, we do this across both mind and body. Which means: the felt sense of our experience matters. So does the context in which we engage it.

Unlike computers that are trained on clean, labeled data, humans learn through constant interaction with a messy, unpredictable world. Our internal systems are always updating—based not just on what we do, but why we do it and where it happens. From an active inference perspective, aligning our intention, activity, and context helps reduce prediction error and create what we'll call ecological coherence. It's a term for when things feel like they click and your actions match your goals in a setting that supports them. In those moments, your system has less to resolve, and more energy becomes available for insight, creativity, or simply presence.

Even small choices can make a difference. Take jogging: a run through a quiet park may support reflection better than one through traffic. Likewise, music with lyrics might crowd your mind, while ambient sounds could leave more room for your own thoughts. These examples highlight how we can use this lens to make gentle shifts toward more coherent states that support learning, regulation, and ease.

The active inference perspective offers us a way to explore this with structure, but not rigidity. It invites us to reflect on what kinds of activity (and in what kind of setting) our system might benefit from in the moment, or that we want to build greater capacity in. Do we need stillness or movement? Mental quiet or cognitive stimulation? A soothing environment, or a challenging one that strengthens our ability to adapt?

**ANNOTATION**

Practical guide inspired by active inference principles, not a strict scientific taxonomy.

To support this exploration and opportunity to update, we propose a two-dimensional model: mental activity (ranging from quiet to active) and physical activity (ranging from still to active). Utilizing this model can help us map the spaces we move through in daily life. Each quadrant offers a different type of update:

- **Quiet mind, quiet body:** Practices like meditation or breathwork train the system to detect subtle signals often drowned out in daily noise. These help reduce stress and support emotional regulation.
- **Active mind, quiet body:** Activities like journaling or reading provide cognitive stimulation without sensory overload and help support symbolic reasoning and internal clarity.
- **Active mind, active body:** Movement that also engages strategy or coordination (like martial arts, dance, or team sports) activates real-time

updating through sensorimotor integration.

• **Quiet mind, active body:** Gentle activities such as walking or swimming allow the mind to soften while the body remains engaged, promoting integration and emergent insight.

The beauty of this approach is its adaptability. You can move within it according to your own needs, goals, and energy. Some days you might inhabit one quadrant. Other days, you might move across all four. This practice functions as a living guide for engagement and is not a strict regimen we need to perfect.

As you become more fluent in navigating these states, another principle becomes important: variation. Research in both algorithmic training and adaptive systems shows that variation, especially when it includes unpredictable or slightly challenging conditions, can improve generalization and resilience. In human terms, this means that engaging with diverse experiences, even those that feel uncertain or uncomfortable, can actually build your capacity to stay grounded and respond skillfully. From an algorithmic perspective, this resembles training under adversarial or perturbed conditions to increase robustness and adaptability.

This applies both across quadrants and within them. For example, within the 'quiet mind, quiet body' space, you might alternate between breathwork, floating, or sound meditation—each introducing different nuances. This kind of intra-quadrant variation functions like training across a diverse but thematically consistent dataset. Over time, these subtle shifts help build more flexible internal models and increase the likelihood of successful updates when life delivers the unexpected. You begin to sense what kinds of activities help you update well according to your intentions. And you gain tools to recalibrate when you drift. In this way, active inference becomes more than a theory. It becomes a practice. One shaped by the rhythms of your own life.

## Adaptation Over Time

Finally, it's worth acknowledging that this process doesn't happen all at once. Our bodies, intentions, and environments are always shifting, and our internal systems adapt gradually in response. One clear example is the way our perception of food evolves over time. Something that once felt bland or unappealing (like unsweetened yogurt at first taste) can come to feel deeply satisfying and nourishing when experienced consistently. Rather than simply being a matter of acquired taste, this illustrates how flexible our predictive systems can be.

Through the lens of active inference, our brains are constantly predicting what a sensory experience will feel like and updating those predictions based on what actually happens. When we repeatedly expose ourselves to nutrient-rich or whole foods, our sensory and emotional responses begin to recalibrate. The taste becomes more rewarding as our systems learn to link it with positive, predictable outcomes, such as stable energy, improved digestion, and a grounded sense of satiation. In short, it's entirely possible to develop new preferences. These preferences, in turn, shape who you are, including the traits and tastes that define you.

These updates typically take shape over days or weeks, depending on the context, history, and individual variability. But they do happen. Over time, the system learns to make more accurate predictions, and with that, our experience becomes more fluent, confident, less reactive, and often more pleasurable. What starts as an intentional shift in input, what we choose to consume, becomes a deeper shift in how we perceive and respond. It's a quiet but powerful reminder that exposure, repetition, and alignment can gradually shape a more coherent internal experience and increase in ease over time.

## SPARK

Consider the four quadrants of mental and physical activity and your overall intentions with your system. You might even sketch the model and label where your current activities land within it and label your overall intentions at the top. Is there an area that feels underexplored or that dedicating more time might better support your intentions? Are there areas that are serving you especially well you'd like to maintain? Or areas that could benefit from a bit more variety or intentional challenge?

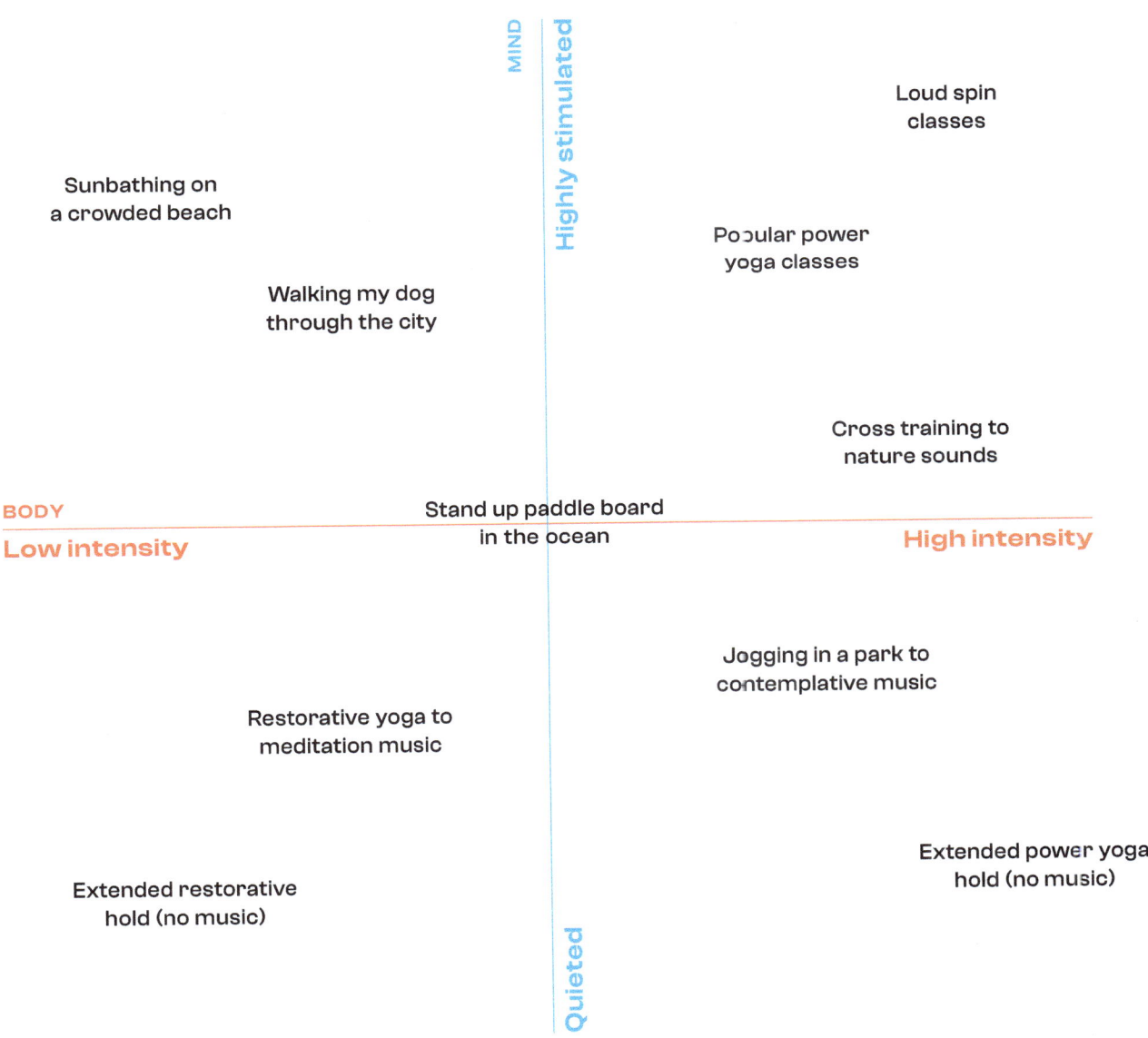

MIND

Highly stimulated

Loud spin
classes

Sunbathing on
a crowded beach

Popular power
yoga classes

Walking my dog
through the city

Cross training to
nature sounds

BODY

Low intensity

Stand up paddle board
in the ocean

High intensity

Jogging in a park to
contemplative music

Restorative yoga to
meditation music

Extended power yoga
hold (no music)

Extended restorative
hold (no music)

Quieted

# REFRAMING "TRUTH": UNDERSTANDING THE LIMITATIONS OF OUR PERCEPTIONS

# REFRAMING "TRUTH": UNDERSTANDING THE LIMITATIONS OF OUR PERCEPTIONS

Now that we've explored how we can shape our active inference cycles through intentional activity, we can begin to deepen our understanding of what those cycles are actually generating. Our resulting perception of life—a blend of body, brain, and surroundings—is, in many ways, a hallucination. Take color, for example. Colors don't exist out in the world; our vibrant experience of them is entirely constructed by the brain interpreting light. It's as if we're small organisms navigating our environment by making sense of limited signals, layering meaning where needed, like adding color, just to survive. No wonder we feel an innate urge to understand what's happening around and within us. Without that drive, survival would be far more difficult.

Today, that desire shows up in many ways, from self-tracking our steps, how we read body language in social settings, or even market forecasting in business. It's deeply embedded in us: our brain is wired to minimize uncertainty and exert control where possible, because doing so has historically increased our chances of survival. But in truth, we can never know with absolute certainty what's happening, only approximate the most likely explanation. Because we're ultimately limited by what we are—self-generating, organic organisms deriving interpretations of what's happening around us.

That's not to say we can't reach high degrees of certainty. Take, for example, my little white dog, who is lying in the sun beside my desk while I write this. While my eyes have never been through a certification process like a piece of lab equipment, they have been checked at an optometrist, and since my birth, my active inference cycle has been refining what I see. Suffice it to say, I can have a very high level of confidence that my dog is indeed napping happily in the sun. We could call this a subjective truth, since it's a truth I've derived using my active inference cycle.

However, this certainty starts to break down when I ask the following question: *Is my perception of the white of my dog's fur the same perception my partner has when he looks at our napping dog?* While the objective sensory input (the white fur) may be the same, the subjective perceptual experience arising from that input could be substantially different. My perception of "white" is filtered through my unique life history, including the particular shades of white I was exposed to growing up, my conceptual understanding of white as a color, and so on. Similarly, my partner brings his own background to bear, different levels of attention, and subtle contextual differences, given he's in a different part of the room than I am. We could take it even a step further. The Inuit, for example, have multiple words for different types of snow, distinguishing subtle variations that an English speaker might not even notice. If they were to magically appear in the same scene, their perceptions would likely be markedly different from mine and my partner's—not just in what they see, but in how they categorize and interpret the scene. For someone accustomed to differentiating countless variations of snow, the specific textures, densities, and hues in my dog's fur would likely stand out in ways I would never think to notice. My expectation of what someone might focus on is shaped by my own experiences, but their attention would be guided by an entirely different perceptual understanding, further varying the differences between our individually derived "truths". **Thus, this example helps illustrate just how complex and subjective our derived truths are, revealing that even in a simple shared moment, different individuals can experience reality in profoundly different ways.**

The lack of "absolute truth" is a bit of a tricky one to wrap one's head around, given that deriving understandings about the world around us is so key for our life experience, not to mention survival. Let's take, for example, your birthday. You likely have a day that immediately pops into mind. You've likely looked at a birth certificate (used your eyes to sense the information and your mind to cognitively understand the day) and have grown up being told by your caregivers that's the day you were born (used your ears to hear their words and then your mind to infer that's the day I was born). However, once again, we can see this process requires passing through your bodily sensory system and your cognitive inference system rather than being based on some magical, verifiable, truth-telling system. The truth about the day you were born is simply an inference based on the best possible explanation for the evidence available to you at the moment.

This means we're limited in the concept of an absolute, objective truth. Instead, we could think of words more like subjective, derived, or individual truth. Our goal is not to find "absolute truth" but rather the best explanation that minimizes free energy (such as doubt, ambiguity, or complexity) and best fits our sensory data. If we think back to our birthday example, imagine discovering that your birth certificate had a clerical error. For years, you believed one date, but upon discovering the mistake, you now have to update your understanding. This illustrates the idea that "all models are wrong, but some are more useful than others." Even if the original date was inaccurate, it served as a useful framework until better evidence became available. Even the corrected clerical mistake has limitations—different calendar systems exist around the world, meaning your birthday could be calculated differently depending on cultural or historical contexts. Additionally, time itself is relative; factors such as time zones and the position of Earth in its orbit mean the exact moment of birth can be perceived in various ways. Even relativity from an astrophysical perspective can mean the moment of birth can have different interpretations.

Similarly, our personal relationships often reveal how our understanding of others changes over time, influenced by new experiences, conversations, and emotional insights, realizing that past perceptions were based on incomplete or biased information. As our perspectives evolve, we see more complexity in others, realizing that earlier judgments often reflected the limits of our own understanding rather than "absolute" truth. Recognizing this allows us to approach relationships with greater insight, openness, and adaptability.

Thus, we can hopefully start to sense and understand that truth is shaped by individual experience and internal models, with each person perceiving the world through their own lens. This also becomes particularly evident in politics, where uncertainty fuels strong emotions like frustration, hope, skepticism, and relief. Rather than getting swept up in these reactions, we can remember that everyone is interpreting the world through their own active inference cycles. It's not always easy—especially when others support policies that feel directly opposed to our values. While taking breaks can be healthy, disengaging entirely prevents us from contributing to collective understanding. We may never reach certainty, but we can refine our perspectives toward more useful models.

Active inference cycle with
**overwhelming levels of information**

Active inference cycles with
**balanced levels of information**

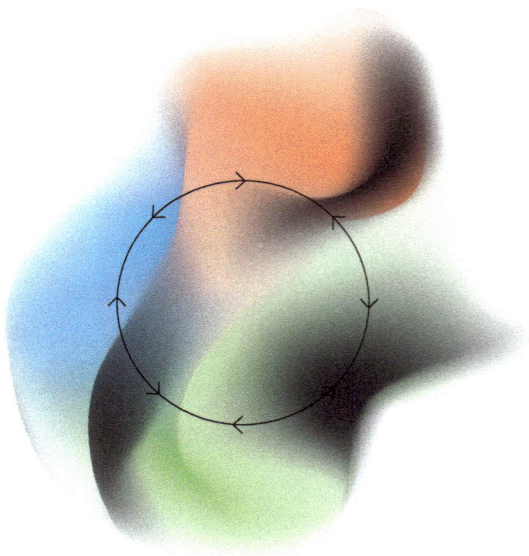

This mindset has become a personal practice for me. I intentionally challenge my assumptions, even in emotionally charged moments, to build resilience in my own active inference cycle. I stay open to perspectives across political and cultural lines—not always comfortably, but consistently. This helps me attune to truths with lower free energy, even when they conflict with dominant narratives. For some, this can be challenging, but it's necessary for a functioning democracy—one that depends on our willingness to engage across differences rather than retreat into isolated realities.

Finally, we can pull back in the bidirectional nature of the brain and our active inference cycles. Remember, our cycles not only absorb what's happening but have an arrow pointed in the opposite direction as well. **What our cycles expect to perceive affects what we perceive**. This means our cognitive load increases when we encounter misleading, contradictory, or high levels of information. Additionally, the effort required to process and regulate challenging emotions in these circumstances increases as well. When this happens, it becomes more and more difficult to distinguish between useful and misleading models of reality.

We spoke earlier about how our energy is finite. If we create environments that flood our senses with less useful information, we can struggle to limit their impact on our active inference cycles. A historical example is the witch trials that happened centuries ago, where thousands of people were accused and executed. Today, the overwhelming majority of us struggle to comprehend how this happened. However, at the time, authoritative-sounding evidence surrounded individuals, making it difficult for many to recognize how flawed the model of witches causing diseases and famine was compared to the more accurate explanations involving microbes and climate patterns. Their sensory information, shaped by prevailing beliefs and limited scientific understanding, reinforced these misconceptions and made it challenging to arrive at more useful interpretations. Recognizing the limits of the human active inference cycle in processing our truths is highly valuable for us to understand, especially as we incorporate more and more computing power into our societies. How will we balance refining and deriving effective understandings of truth without overwhelming our systems?

## SPARK

✦ Consider how you might support others knowing they're navigating the world through their own self-generating models—shaped by their bodies, minds, and life experiences. What might it sound like to connect in a way that respects these differences? Could it mean asking thoughtful questions instead of offering directive advice? Or giving space for their truths, even when they differ from yours? Perhaps it's an invitation to grow together by cross-pollinating perspectives and updating your own understanding in the process?

**Navigating Uncertainty**

"We can never know with absolute certainty what's happening, only approximate the most likely explanation.

# THE SCIENCE WITHIN: BELIEVING IN OUR BUILT-IN CAPABILITIES

# THE SCIENCE WITHIN: BELIEVING IN OUR BUILT-IN CAPABILITIES

We just told you there is no such thing as final or absolute truth, but we're writing and creating a book that tells you active inference and the free energy principle are useful for you to understand and navigate life with. If you're like many, it might feel a bit discombobulating. This section is meant to help.

## Robust Science

Science, as a word and as a concept, has been under increasing scrutiny in recent years. I like to use the term "robust science" to describe a class of scientific inquiry that builds confidence through rigorous validation and replication across different institutions and perspectives. Science, at its core, is a systematic approach to understanding the world through observation, experimentation, and refinement of theories based on evidence. Within the framework of active inference, science helps refine our models of the world by minimizing uncertainty and updating our beliefs through new data.

Using this approach does not mean that we can automatically develop beliefs in a single finding. Consider the example of a stand-alone paper that claimed coffee consumption reduces the risk of a specific disease, only to be contradicted by later studies showing no such effect. However, when multiple studies build on one another (such as through meta-analyses that aggregate findings across institutions and methodologies), we can infer higher levels of confidence. Active inference has been supported by several such studies. For example, review syntheses have shown its relevance across neuroscience, artificial intelligence, and psychology.

System thinking further validates active inference by examining its applications across domains. From healthcare to robotics, active inference principles have been successfully applied, reinforcing its robustness. Karl Friston, the pioneer of this framework and key advisor of this book, is among the most cited researchers in the world, with citations surpassing those of Einstein and Hawking. This widespread early application across institutions, cultures, and disciplines signals the robustness of this model. Many modern industries are already applying principles derived from active inference, including artificial intelligence, autonomous systems, and cognitive computing.

As a research scientist, I find it very compelling when a framework demonstrates strong support across diverse applications, which served as an early indicator of active inference's potential for me. The robust science behind active inference suggests it is a highly valuable model, yet we remain aware of balancing our confidence. While those of us contributing to this book hold active inference in high regard, we recognize that our understanding needs to evolve alongside ongoing scientific advancements. Two things we can state with high confidence: active inference is supported by robust science that can help reduce uncertainty and dissonance in our lives, and our knowledge will continue to grow and refine as new research and perspectives emerge.

## Our Physiology: A Remarkable Support System

Another lens through which we can view active inference is our physiology and how we are wired to support this process. The active inference cycle begins with sensing our environment. Our bodies possess numerous highly effective systems to achieve this, including the traditional five senses: sight (visual system), hearing (auditory system), touch (somatosensory system), smell (olfactory system), and taste (gustatory system). In addition, our body sensing includes lesser-known systems balance (vestibular system), body position and movement (proprioceptive system), pain detection (nociceptive system), and temperature sensing (thermoceptive system), all working together to support our perception. Put simply, our body efficiently carries out a vast amount of sensory work, automatically operating largely outside of our conscious awareness to support active inference.

Once sensed, these signals are processed through various physiological systems. The nervous system plays a crucial role, with the central nervous system (CNS) integrating sensory data with past experiences to form perceptions, and the peripheral nervous system (PNS) transmitting sensory information to the brain for interpretation. The autonomic nervous system (ANS) regulates involuntary responses, balancing the sympathetic (fight-or-flight) and parasympathetic (rest-and-digest) responses to optimize perception and readiness. The limbic system, including the amygdala and hippocampus, attaches emotional significance to sensory experiences, shaping how we perceive the world.

**ANNOTATION**

Sometimes, I like to think of these as 'science chakras'—systems I gradually learn over time, not by memorizing a list, but by understanding my experiences and recognizing which systems are likely reacting. This awareness increases my self-knowing, helping me to respond more effectively.

**DEFINITION**

A term sometimes used
to describe the brain's
distributed mechanisms
for perceiving and
estimating time.

Additionally, time perception and temporal processing systems help track the passage of time, allowing us to anticipate and plan actions effectively. Emotional and social processing systems, such as the prefrontal cortex and the mirror-like neuron system, help us interpret social cues and regulate our responses, facilitating interactions and decision-making in complex social environments. These intricate systems work naturally in the background, effectively supporting our active inference cycles, allowing us to navigate the world with remarkable efficiency.

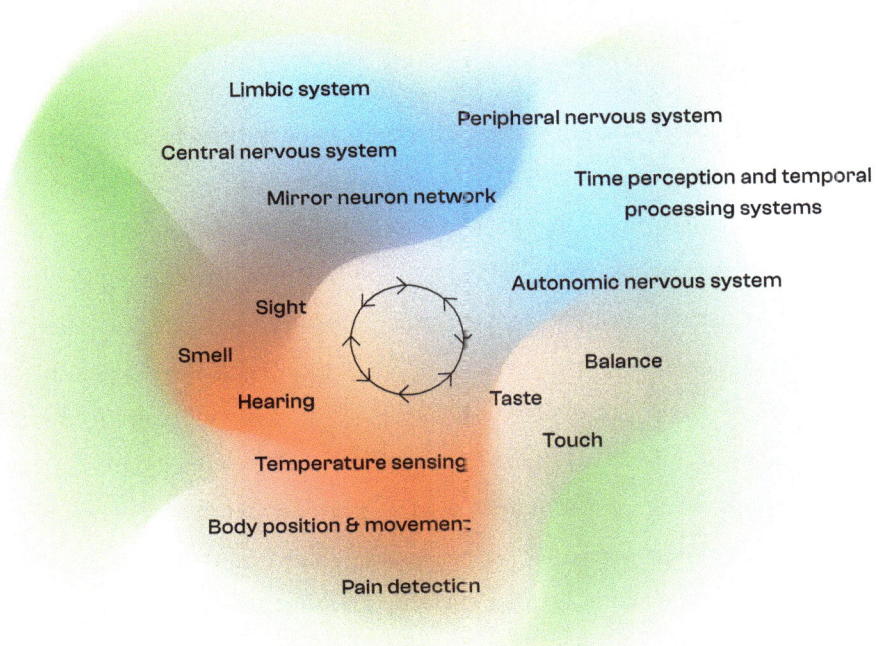

Beyond sensing the external world and processing external signals, our body monitors its internal state through interoception, the processing of its own internal signals. This process is supported by systems such as the cardiovascular system (tracking heart rate and blood pressure), the respiratory system (monitoring oxygen and carbon dioxide levels), and the endocrine system (hormonal regulation). The insular cortex integrates these internal signals, creating a cohesive awareness of bodily states that guide decision-making and action. These processes operate and support us continuously, helping us maintain equilibrium and prepare to respond to the multi-layered shifts within our active inference cycles.

## Beyond Awareness: Believing in Our Inherent Processes

This may or may not feel like a lot of science to you. For some, understanding the science brings clarity, while for others, it may feel abstract or distant. Regardless of how we engage with it, one of the most empowering realizations is that we are not separate from the process—we are an intrinsic part of it. We can be understood as manifestations of this process itself. Just as breathing sustains our physical existence often without thought, active inference continuously unfolds within us often without our conscious realization, supporting our perceptions, behaviors, and interactions with the world. In Chapter 19, we explored the aspects of this process we can consciously influence. It's reassuring to recognize that we are dynamic expressions of active inference, inherently equipped to navigate life with remarkable capability and resilience.

**ANNOTATION**

Different manifestations of active inference prefer to make their inferences in different ways. If deep maths and science is your way, I might suggest joining the Active Inference Institute or attending the International Workshop on Active Inference

**Navigating Uncertainty**

## SPARK

Think about the ways you've come to recognize your system supports you in your active inference cycles, similar to how your breathing sustains you, often without conscious effort. Make a list of the ways your cycle helps you adjust, respond, and make meaningful shifts, perhaps often without your direct awareness. Consider keeping this list somewhere meaningful, and revisit or add to it as your knowledge deepens.

# HOPEFUL & NOTEWORTHY APPLICATIONS

# HOPEFUL & NOTEWORTHY APPLICATIONS

As we've seen throughout the book, the science we've used to better understand ourselves and our physiological capabilities can also be applied in broader, real-world applications. Imagine a world where technology embodies the principles we've been exploring, such as adaptability, efficiency, resilience, and harmony, and creates novel solutions to humanity's most pressing challenges. While the idea of technology so deeply embedded in everyday experiences might initially feel frightening, envisioning such a future, done in ways that consider these deeper themes, can also inspire hope. Active inference and the free energy principle are already helping to make this vision a reality, opening doors to transformative applications in diverse and unexpected areas. Here, we'll dive into three particularly exciting and hopeful examples that illuminate the meaningful impact these principles can have on our world.

## Verses: Eco-friendly Artificial Intelligence

Artificial Intelligence has often been critiqued for its power-hungry nature. Large language models, autonomous vehicles, and deep learning networks usually require immense computational resources, resulting in substantial environmental footprints and power costs. Verses takes a novel approach, applying active

**ANNOTATION**

This aligns closely with how I view artificial intelligence: Are we deploying it in ways that preserve and enhance our human experience—supporting our bodies as sensory organs, fostering meaningful dialogue instead of fragmentation, and creating opportunities that benefit many rather than just a privileged few?

inference principles to enhance AI efficiency and sustainability.

Verses aims to use active inference's predictive capabilities to optimize how AI processes information, significantly reducing computational demands. Rather than constantly analyzing vast data streams, systems based on active inference carefully select information that resolve their uncertainty, drastically cutting down unnecessary processing. This approach turns AI search engines into smart data miners and can significantly cut energy use.

Imagine autonomous vehicles or smart cities powered by this technology, consuming less energy, minimizing their carbon footprint, and significantly lowering power costs. The hope here is a cleaner, affordable future without compromising technological advancement.

## Cortical Labs: Biological Computing Made Accessible

At Cortical Labs, the blend between biological neurons and digital computing is becoming a reality. Their "DishBrain" project is an audacious yet elegant demonstration of active inference in action, using real neurons to learn, adapt, and perform computational tasks. DishBrain employs a fusion of biological neurons and silicon interfaces, harnessing active inference principles to enable living neurons to predict and respond to digital stimuli.

**ANNOTATION**

Such projects also raise questions about how we relate to biological life forms, even at the cellular level.

The implications of DishBrain are enormous. Such bio-computational devices offer the potential for remarkable adaptability, learning efficiency, and resilience compared to traditional silicon-only technologies. For instance, biomimetic-computational devices like DishBrain, neuromorphic computing, photonics, and memristors could revolutionize personalized medicine by dynamically adapting to an individual's unique biological responses, optimizing treatments in real-time.

What makes Cortical Labs especially intriguing, however, is their commitment to an open-source model. The open source model strategically leverages collective ingenuity and has the potential to accelerate progress in neuro-inspired technology while distributing its benefits. In doing this, collective insight becomes a strength to all involved.

## As a Tool for Governance

As we've explored throughout this book, active inference underscores the intricate interplay between mind, body, and environment, highlighting the profound influence our emotions, bodily sensations, and thoughts have on our behaviors. However, computational methods grounded in active inference could systematically incorporate affect-like signals, analyzing them objectively rather than reacting impulsively as humans often do under stress. This approach allows the models to maintain greater equanimity and clarity, even in emotionally charged scenarios. **When analytical tools are used alongside human emotional and bodily insight, governance has the potential to become significantly strengthened, especially in emotionally charged decision making scenarios.**

Consider the critical decisions following the September 11th attacks in the United States. These decisions led to prolonged military engagements where

**ANNOTATION**

Akin to emotional and bodily sensations.

anticipated weapons of mass destruction were ultimately not found, contributing to substantial financial debt and tragic loss of life. Reflecting on footage from that period, we see individuals experiencing understandable emotional reactions amid extreme uncertainty. While opinions vary on the wisdom of the choices made, it is notable (and gently reassuring) that leaders managed the situation without even more catastrophic outcomes like escalating into a nuclear war.

Regardless of one's political views, we might collectively ponder: what would governance have looked like if leaders had access to more robust, computational tools capable of modeling likely outcomes accurately? Could more precise predictions regarding the existence of weapons of mass destruction have changed the decision to initiate war? Might a clearer assessment of probable long-term consequences have prevented significant expenditures, both monetary and human?

As we've looked at previously, under stress, humans often experience narrowed perception and cognitive rigidity, impairing clear decision-making. Computational models leveraging free energy principles, aligned with our physiological and ethical frameworks, could substantially clarify these critical moments. Crucially, however, AI alone cannot dictate ethical choices—it needs to be guided by human-defined values. These values could be determined in calm, reflective states, ensuring that computational tools provide meaningful clarity precisely when we most need it.

## Computational Psychiatry: Personalized Paths to Healing

This application strikes a particularly resonant chord for me. Like many, I have personally navigated the often opaque journey toward healing and behavioral change, struggling to find actionable and timely insights into my own psychology. Computational psychiatry, informed deeply by active inference and the free energy principle, offers profound hope.

By modeling an individual's cognitive patterns and psychological states through active inference frameworks, computational psychiatry can provide highly personalized and targeted therapeutic interventions. Imagine clinicians equipped with predictive models capable of identifying precisely when and where interventions can most effectively shift maladaptive cycles or reinforce positive behaviors. Treatment becomes less about guesswork and more about precise, personalized care.

This targeted approach signals a breakthrough in mental health treatments, from anxiety and depression to more complex psychological conditions. When therapists work through the lens of active inference, they're better able to align treatment paths that reflect each patient's individual experience. This approach supports more meaningful progress and provides patients with clearer routes of genuine transformation. **For me, and for countless others who have experienced or witnessed others we care about feeling stuck in their psychological patterns, computational psychiatry inspired by active inference and free energy is remarkably hopeful.**

Together, these examples highlight the transformative opportunity of active inference beyond the focused application of this book. From sustainable and economically accessible technology and biological computing that supports human

**ANNOTATION**

Since our physiology is deeply influenced by the presence and regulation of others, humans will likely continue to play a critical role in this dynamic. Co-regulation is something AI is far from fully replicating. As such, it's likely that human spiritual leaders, coaches, and therapists will remain essential, even as their roles evolve alongside advancing technologies.

**ANNOTATION**

Governance is another domain where active inference could be a useful resource, helping institutions function more like adaptive systems— sensing, predicting, and dynamically responding to the evolving needs of their people. In doing so, they become more resilient and attuned.

**Navigating Uncertainty**

needs rather than hindering them, to deeply personalized and more effective mental health treatment, these examples illustrate the hopeful potential and opportunity of active inference cycle and the free energy principle on a larger scale. They are a key method we can utilize to shape technology to meet our deeper human needs and create unprecedented opportunities for life.

# GETTING CREATIVE: CARRYING THIS FORWARD

# GETTING CREATIVE:
# CARRYING THIS FORWARD

Learning new concepts takes time and energy. If you've made it this far, you haven't just read about science or self-understanding. You've already begun the process of integrating it.

These final sections offer a handful of distilled insights paired with small, flexible practices for the mind, body, and environment. They're not prescriptions or rigid routines, nor is it likely feasible to do all of them at once. Rather, they are meant as inspirations and a resource: starting points to deepen ease, adaptation, and resilience in your own life and perhaps refer to over time.

As you read through them, consider pausing and noticing what stands out. Which practices feel resonant? Which ones might gently stretch you? What do you have space for at the moment and what might be better for another time? Which might not be right for you? For many, choosing just one for mind, body, and environment is likely enough. Tuning into your system can help illuminate what fits. And if it feels right, consider setting a soft intention (or even a calendar nudge) to revisit this list in the future. What resonates with you may shift over time. That's part of the beauty of this process: it shifts as you do.

## 1. You're an Active Inference Cycle

You're not separate from the process. You're a living expression of it. Life moves through you in every moment. You have the beautiful opportunity of, and are ultimately shaped by, the gift of life your cycle manifests. Your body is designed to help you navigate uncertainty—your sensory systems, brain, and physiology engage with it rather than being passive observers. They adjust and respond continually, supporting you as conditions change

### Activity Inspirations

**REFLECTIVE JOURNALING:** consider prompts that might help surface subtle dissonance and free energy or foster greater coherence across your system and active inference cycle. Some examples include: *"What area of my life feels slightly 'off' or misaligned? What might my system be trying to communicate?"* or *"If I tuned into my system right now—without trying to fix anything—what might I notice?"*

**BREATHWORK:** consider breathwork that might serve your system or help soften underlying tension. Is it brief and grounding, like a few conscious, steady breaths? Or is it more immersive and exploratory, like engaging in a guided session of intense breathwork with a trained practitioner?

**NATURE IMMERSION:** consider getting out into nature in a way that supports your system or reveals subtle dissonance worth exploring. It might be as simple as a walk through a nearby park, or a more immersive experience like hiking outside the city. For some, this might mean working with a trained practitioner in a guided, nature-based microdosing session, providing grounding and insight within a safe and intentional setting.

## 2. Mind, Body, and Environment Are Always in Dialogue

You are not a separate observer of these parts. You are the system they create together. When we notice and nurture their interplay, we invite a deeper coherence that supports adaptation, expression, and well-being. Even subtle changes in one domain can ripple across the others. A gentle breath, a supportive conversation, or a small mindset shift can recalibrate the entire cycle.

### Activity Inspirations

**CONCEPT MAPPING:** A visual method for organizing ideas and making relationships between them visible. Start by writing key concepts (like "mind," "body," or "environment") on a page. Then draw arrows or lines between them, adding short notes that explain how you notice one influence, support, or interact with another.

**STRETCHING OR YOGA:** Try choosing a few slow, mindful movements (like a gentle twist, forward fold, or shoulder roll) and notice how your body feels before, during, and after.

**CURATING WORKSPACE FOR SENSORY EASE:** Take a few minutes to look around your workspace and notice what feels calming or overstimulating. Try removing visual clutter, adjusting lighting or sound, or adding textures or objects that help your system settle.

### 3. From Dissonance to Direction, Balancing Free Energy

Free energy arises when expectations clash with reality. These mismatches serve as signals rather than faults. They offer opportunities to pause, reflect, and adjust. Because ultimately, your system is built for this. Your body and brain are built to respond to mismatches and are continuously recalibrating. Each moment of dissonance provides opportunity for learning, flexibility, and growth, orienting the system toward deeper coherence.

**Activity Inspirations**

THOUGHT LABELING: Briefly notice and name the type of thought you're having (e.g., worry, planning, judging) to create a bit of space between you and the thought.

PROGRESSIVE MUSCLE RELAXATION: Gently tense and then release one muscle group at a time (from your toes to your face) to support body awareness and relaxation.

ENGAGING IN COMMUNITY DISCUSSION: Share your perspective and listen to others in a trusted group or space, allowing mutual insight and reflection to unfold.

### 4. Flexible Focus: Adjusting the Lens Within

Practice adjusting how much confidence you place in thoughts, sensations, and perceptions. Explore zoomed-in focus with wide-angle awareness. When we assign too much certainty to fleeting emotions or automatic thoughts, we increase dissonance and tension in the system. When we can widen our view or soften focus when needed, we navigate uncertainty with more clarity and ease.

## Activity Inspiration

**SENSORY DEPRIVATION TANK:** Spend time in a quiet, dark float tank to reduce sensory input and notice how your attention and awareness shift in the absence of external stimuli.

**BODY SCAN MEDITATION:** Slowly bring attention to different parts of your body (from head to toe), pausing at each one to notice sensation without judgment.

**NOTICING MIND-BODY-ENVIRONMENT SHIFTS IN A NEW SPACE:** Step into a different environment and gently observe any shifts in your thoughts, bodily sensations, or energy. What feels different, and what might that reveal?

## 5. The Body As a Sensory Organ

The body holds its own kind of sensemaking. Tuning into that felt sense can help increase clarity, capacity, and ease. The brain constantly monitors signals like heartbeat, breath, and gut tension to infer how things are going. That felt sense is information.

## Activity Inspiration

**SOCRATIC QUESTIONING:** Ask yourself open-ended questions like *"What else might be true?"* or *"Is this just a familiar thought pattern?"* to explore and gently challenge assumptions.

**Navigating Uncertainty**

**BALANCE TRAINING:** Stand on one leg or shift weight slowly from side to side or practice using a balance board to build awareness of your body's position and steadiness.

**SENSORY CALIBRATION IN DIFFERENT ENVIRONMENTS:** Visit different spaces (like a quiet room, a park, or a loud communal gathering) and notice how your body, thoughts, or energy shift in response.

## 6. Play Is Often Productive (And Fun!)

Joyful exploration, far from a frivolous luxury, offers a highly effective way to reduce uncertainty and learn what's possible. It helps systems recover from stress, expand tolerance windows, and develop creative responses to novel situations. Group play, in particular, supports co-regulation, builds trust, and lowers shared uncertainty, benefiting not just individuals but whole groups.

### Activity Inspiration

**BRAINSTORM A SILLY IDEA:** Set a brief timer and come up with outlandish or playful ideas, no judgments.

**EXPLORE SENSORY PLAY:** Engage with textures, sounds, or movement that feel good, like playing with putty or playing with different textures across another's skin.

**INITIATE PLAY IN A SHARED SPACE:** Invite others to join in a light, spontaneous activity, such as a game or a funny challenge.

**ANNOTATION**

A workspace favorite of mine is inciting nerf "battles" or kind prank culture.

## 7. Volition Over Willpower

Willpower relies on top-down override; volition arises from system-wide integration. When your body, environment, and values are in sync, you don't need to push in the same way, and effort softens. Actions arise from coherence, grounded in your experiences, current needs, and unfolding sense of direction.

### Activity Inspiration

INTENTION REFRAMING: Gently revisit a goal or action with curiosity. Are there updates from within or in your environment that might help guide a more adaptive next step?

TACTILE ANCHORING: Use a small physical object or texture (like a smooth stone or fabric) as a sensory cue to return to presence or intention.

VOLITIONALLY INSPIRING BOOKS: Keep a few books nearby that reconnect you to purpose, insight, or creativity. Revisit them when direction feels unclear.

## 8. Rest Can Clarify the Signal

Stepping back isn't stepping out, and rest is far from passive. Rather, it's an active and essential part of how your system processes the world. It allows your system to sort signal from noise, discern what truly matters, and recalibrate without the overwhelm of constant input. **You don't stop learning when you pause**. Interestingly, some of the deepest insights can arise in stillness; when attention softens, internal patterns reorganize, and clarity has the space to surface.

## Activity Inspiration

**WINDOW GAZING:** Gaze softly out a window for a few minutes without trying to focus on anything in particular. Let your mind and eyes rest.

**RESTFUL BODY POSTURE HOLDS:** Try a posture like lying down with legs up the wall, or reclining with support. Stay for a few or several minutes and notice any shifts in how you feel.

**VISITING PLACES THAT ACCOMMODATE REST:** Go to a location designed for rest, such as a library, quiet garden, or calming café, and allow yourself to simply be there without an agenda.

## 9. Self-Knowing Is a Continual Illumination

**Understanding your system reduces unnecessary free energy.** When you can anticipate your own tendencies, triggers, and needs, your system doesn't waste energy managing surprise. Instead, it can reallocate that energy toward growth, recalibration, or intentional response. The more clearly you recognize your patterns, the more choice you have in shaping your path. Insight and integration unfold through reflective action where awareness meets curiosity and small shifts can ripple outward.

## Activity Inspiration

**DREAM JOURNALING:** Keep a notebook by your bed and jot down any dreams upon waking, capturing sensations, themes, or images before they fade.

**WALKING MEDITATION:** Walk slowly and with intention, paying attention to the sensations in your feet, breath, and surroundings.

**VISITING A MEANINGFUL LOCATION:** Spend time in a place that holds personal significance, and gently notice what thoughts, feelings, or memories arise.

## 10. Intention Shapes Experience

**Intentions act like predictive models.** They influence how your system processes information and prepares for action—shaping what you notice, how you interpret it, and what you do next. The world doesn't just act upon you— you shape it, too, in a continuous feedback loop. When your internal models are aligned with your values, bodily signals, and external context, even small intentions can meaningfully shift your perception and behavior. Through the lens of active inference, setting intention helps reduce prediction error, promoting clarity and coherence across your system. In doing so, imagined possibilities become more achievable outcomes—rooted in physiology, shaped by experience, and guided by adaptive response.

### Activity Inspiration

**INTENTION SETTING:** Pause and clearly name what matters most to you in this moment, setting a gentle guide for your attention and actions.

**INTENTIONAL BREATH-TO-MOTION ACTIVITY:** Inhale and roll your shoulders up; exhale and roll them back and down.

Experiment with muting, unfollowing, spending less, or more intentional time on social media. What arises?

## 11. Safety Creates Space to Soften and Shift

**Safety often reduces free energy.** When your system no longer has to defend against or predict danger, it frees up resources for connection, learning, and growth. Feeling safe (both internally and in relationships) helps the system soften. That softening opens the door for curiosity, fresh perspective, and meaningful change to emerge. But safety isn't always available. One possible response is **naming the lack of safety without judgment**. Simply saying, "I don't feel safe right now," can begin to reengage the prefrontal cortex and soften reactive patterns. Another response is to seek partial safety. Even small anchors, such as a steady breath, a grounding texture, a calm environment, or the steady presence of someone you trust, can signal enough safety for the system to begin settling. Over time, these micro-signals can build capacity and open the path for deeper regulation and resilience.

### Activity Inspiration

**SELF-COMPASSION MEDITATION:** Sit quietly and bring to mind a gentle phrase (like *"May I hold myself with compassion"*) while noticing any sensations or emotions that arise.

**PARTNERED BREATH SYNCHRONIZATION:** Sit with a trusted partner and gently match the rhythm of your breath with theirs, perhaps while sitting intertwined, to support connection and calm.

**PRACTICING BOUNDARIES IN CONVERSATION:** During dialogues, notice your comfort level and practice kindly stating a need, limit, or request that supports your well-being.

## 12. Uncertainty Is Fertile Ground

**Free energy isn't always a problem.** Sometimes, it's what sparks transformation and fuels curiosity, movement, and entirely new ways of seeing. Uncertainty can serve as a signal that it's time to update our internal models, rather than something to eliminate. **Letting go of certainty supports flexibility.** It allows the system to revise predictions and avoid the rigidity that blocks perception and growth. Not everything can be known—and that's not a flaw in the system, but a feature of how we evolve. When we allow space for the unknown, we make room for new priors to emerge, creative adaptations to form, and deeper insights to take root.

### Activity Inspiration

**ENGAGE WITH PARADOX:** Sit with a question or tension that doesn't have a clear answer, and allow yourself to notice what thoughts, feelings, or insights arise without forcing resolution.

**COLD PLUNGE:** Briefly immerse your body in cold water or an ice bath to practice sitting with an activated nervous system—notice your breath, sensations, and how your state shifts afterward.

**VISIT AN ABSTRACT ART MUSEUM:** Wander through abstract art and experience it noticing what arises within you.

## 13. Love Often Broadens Possibility

**Connection enhances prediction accuracy.** Being in close, trusted relationships helps regulate each person's nervous system and sharpens the ability to interpret social and environmental cues accurately. Additionally, various forms of connection support adaptation. Relationships rooted in trust and empathy allow systems (including yours) to grow in unexpected and beautiful ways.

### Activity Inspiration

**LOVING-KINDNESS MEDITATION:** Silently repeat simple phrases like "May I be safe, may I be well," first toward yourself, then toward others.

**TOUCH (SELF OR WITH OTHERS):** Gently place a hand on your heart or place your hand on someone else's, noticing the comfort or connection it brings.

**VOLUNTEERING OR SUPPORT GESTURE:** Offer help, encouragement, or time to someone else in a way that feels fitting.

## 14. Neurodiversity As a Collective Asset

Your unique way of perceiving and creating has value. Groups that embrace a variety of internal models tend to explore more possibilities, tolerate more uncertainty, and generate more innovative solutions. Our collective growth comes from reaching toward one another rather than trying to erase differences.

This process deepens shared understanding and enriches the insights we create together.

> **Activity inspiration**
>
> **COGNITIVE DIVERSITY REFLECTION:** Briefly reflect on how different thinking styles, yours and others', have brought value to your experience.
>
> **INTERACTIVE SPORTS:** Join in a sport or movement activity that requires responding to others in real time, like soccer or dance, and notice how it affects your attention and connection.
>
> **WORKSPACE CUSTOMIZATION:** Adjust your workspace to reflect your preferences and needs. Try adjusting screen or chair height or desk arrangement to support better posture, focus, and ease in your body and mind.

## 15. Truth Is Provisional —But Deeply Valuable

**DEFINITION**

Your models of the world will never be perfect, but they can be refined. All understandings approximate reality—some just approximate it better. Ultimately, the brain doesn't aim for objective truth, but for workable interpretations that minimize surprise and guide effective action.

Existing as our best understanding at this moment—open to refinement and evolution as new insights or experiences emerge

> **Activity inspiration**
>
> **REFLECTIVE JOURNALING ON PERSONAL ASSUMPTIONS:** Set aside a few minutes to explore questions around key ideas like, *"When did I first start thinking this, and has anything changed since then?"* or *"Where might my assumptions benefit from an update?"*

**CONTACT IMPROVISATION:** Join a class or explore gentle movement with a partner, staying present and responsive to each other's touch, pressure, and direction.

**VISITING PLACES THAT STRETCH YOUR COMFORT ZONE:** Spend time in a new or unfamiliar setting, like a different cultural space, social environment, or part of town, and observe how it impacts your thoughts, body, or energy.

## 16. You Are the Author of Your Cycle

**You can participate consciously in shaping what's next**. Tuning into your system with intention allows you to harmonize prediction with possibility, turning everyday actions into acts of authorship. You get to choose which inputs to tune into, which of your brushes to use, and what kind of masterpiece your life becomes. No piece will be the same. Yours is a unique and beautiful opportunity.

### Activity inspiration

**CREATIVE VISUALIZATION:** Close your eyes and picture a moment, place, or future that feels aligned with your values. Notice the images, emotions, or sensations that arise.

**KINESTHETIC EXPRESSION:** Use movement (like dancing, stretching, or swaying) to express how you feel without needing words.

**SHARING PERSONAL TRUTH IN A TRUSTED SPACE:** Speak honestly with someone you trust about what feels real for you, allowing connection and reflection to unfold.

These are just a few ideas to get you started. Some personal favorites of mine include slow jogging through my neighborhood park while listening to a Yielding playlist by my friend Jeff Krone, unwinding in the evening with a book by Elizabeth Strout or a Kurzgesagt video, and exploring insights on active inference from Karl Friston, Andy Clark, Anil Seth, and Lisa Feldman Barrett while working on a puzzle. I also maintain intermittent silent meditation retreats and group practices, and regularly sprinkle in new explorations as opportunities arise. However, each life, each work of art, is unique. Ultimately, this serves as an invitation to get creative, drawing inspiration from scientific insights as you paint your path forward.

Nature immersio

Volitionally inspiring books

Engaging in com
discussion

Initiate play in
a shared space

Sensory calibration in
different environments

Noticing mind-body-environment
shifts in a new space

Visit an abstract
art museum

Visiting places that
accommodate rest

Changing social
media habits

Practicing boundaries
in conversation

Visiting places that stretch
your comfort zone

Volunteering or
support gesture

Sharing personal truth
in a trusted space

Workspace customization

Walking meditation

Intentional
breath-to-motion activity

Visiting a meaningful
location

Socratic questioning

Cognitive diversity
reflection

Window gazing

ting workspace
sensory ease

Brainstorm a silly idea

Loving-kindness
meditation

Creative visualization

Dream journaling

Concept mapping

Self-compassion
meditation

Intention reframing

Reflective journaling on
personal assumptions

Reflective journaling

Engage with paradox

Intention setting

Sensory deprivation tank

Thought labeling

Progressive muscle
relaxation

Body scan meditation

Explore sensory play

Balance training

Cold plunge

Tactile anchoring

Partnered breath
synchronization

Contact improvisation

Breathwork

Restful body posture holds

Touch (self or with others)

Interactive sports

Kinesthetic expression

Stretching or yoga

317

# SELF-KNOWING: LITERALLY

# SELF-KNOWING: LITERALLY

While practices help shape the flow of our daily lives, self-knowing helps refine the predictive lens through which we interpret them. As I explored the themes of this book, one of the most illuminating realizations for me was understanding how early in life, during challenging or painful experiences, I instinctively distanced myself from uncomfortable sensations—both emotionally and physically. For me, this manifested as a numbing around my heart area, a subtle detachment from sensations that many others seemed naturally tuned into. My brain, striving to predict and control uncertainty, had learned to block out these signals to create a feeling of safety.

Learning the science behind what was happening was only a portion of the updating that was needed. Exploring these messy bodily sensations, even though ambiguous, became critical. They provided rich, nuanced information about my present reality, enabling a deeper coherence between my internal expectations and external experiences. Mentalizing these sensations by actively naming and engaging with them allowed me to cultivate a more flexible, authentic predictive model as well as an ability to start interpreting what they were signaling to me. It's my hope that this book, along with the practices and resources included, can help you tune into and work with your own bodily sensations and signals. This can support the ongoing process of updating your internal model and lowering expected free energy in ways that serve your particular story.

Writing this book has illuminated my own self-knowing in ways I couldn't have anticipated. It reached deeper than any coaching, meditation retreat, or coursework—though each of these played its essential part throughout the process. The act of researching and writing itself, driven by a scientific thesis, uncovered hidden currents driving stale looping patterns. Bringing these patterns into awareness can help us move through a process of recognition, healing, and ultimately release of maladaptive loops that hold us back. Ultimately, while this book is inspired by science, the journey has vividly illustrated that **stepping into uncertainty and ambiguity and spaces not always clearly defined by science is a key component of where meaningful and transformative growth unfolds.** It is precisely this brave exploration that opens us to fuller, richer lives, aligned deeply with our physiology and consciously chosen intentions.

Ultimately, to understand my own patterns, I had to look back. People are often surprised when I share that both sides of my family carry histories shaped by deep hardship—one by extreme poverty and political repression in mid-century Portugal, the other by childhood sexual abuse in the U.S. These experiences left long shadows in my own life, embedding themselves into the patterns I later came to understand through the lens of active inference and free energy. Writing this book became a process of surfacing and updating deeply learned expectations and family-patterned ways of coping. It was guided by both science and lived experience, and has offered me a faster path to healing than I imagined possible.

Discussing trauma openly is challenging, given the stigma it carries. Yet precisely because it's so difficult, this book exists: born from my journey toward healing and my determination to offer others similar pathways. We are among the first to be able to so robustly intersect science with personal experience, providing deeper and more effective pathways for transformation. The act of birth itself can be experienced as traumatic, a profound adjustment to life's uncertainties. Choosing intentional ways to shift the ripples in ourselves from previous generations (who were doing their best under their circumstances and resources) can support breaking patterns that might otherwise repeat across generations, allowing for more ease, adaptability, and resilience across time.

These realizations have brought challenging questions into sharp focus; however, since there's less free energy in my system, it also opens up opportunities for clearer, more effective navigation: How have I unintentionally harmed due to unresolved ripples moving through me? How do I love individuals who have harmed me and still persist in hurtful behaviors? Writing this book equipped me not with universal solutions, but with greater self-knowing and resources to ultimately recognize and respond more effectively to previously overwhelming and painful questions. We hope it does the same for you.

I've also realized that overcoming significant hardship can cultivate a unique and powerful resilience. Witnessing remarkable transformations—such as a board member channeling his grief from his son's overdose into creating

**ANNOTATION**

To me, offering this possibility of transformation across generations feels like a deeply meaningful and hopeful part of this work.

preventative resources or a woman transcending challenging origins to become a respected spiritual leader—has deeply inspired me. Perhaps my most profound moment was confronting my childhood misconceptions about love during a silent meditation retreat, experiencing the revelation that what I'd known as love was something else entirely. This recognition, incredibly painful yet liberating, clarified bodily signals I'd misunderstood for years, empowering me to hold and transform intense emotions into genuine self-knowing and purposeful action. Perhaps this is one of the deepest gifts of all—the ability to recognize the value of this process, not only within myself, but in others. **In the end, overcoming hardship can build an incredibly valuable life skill, useful not just for personal well-being, but also as a beneficial asset in work culture, innovation, and genuine connection.**

## SPARK

✦ If it resonates, consider writing a part of your own life story through the lens of active inference. This could be a single event described in one paragraph, or a longer narrative that weaves your personal experience together with insights from psychology, physiology, active inference, and free energy.

"In the end, overcoming hardship can build an incredibly valuable life skill, useful not just for personal well-being, but also as a beneficial asset in work culture, innovation, and genuine connection.

# OPPORTUNITY FOR ACTION

# OPPORTUNITY FOR ACTION

As we close, our hope is that this book has offered not only insight, but a seed. A seed of possibility, of self-knowing, of inner coherence. Maybe you're considering a new habit. Maybe you've begun to see your inner world, or the world around you, with more clarity. Maybe you've simply felt a release, like a sense of a little less free energy in your system, and a little more space for what's next. Or perhaps, your system is feeling ready for a larger leap. Here are a few closing thoughts to support you, wherever your next step may lead.

## Inaction is a Form of Action

We've explored how our systems are constantly adapting; sometimes through movement, sometimes through stillness. What looks like inaction is sometimes a deliberate response that offers preservation, integration, or a chance to recalibrate. Just as rest can refine our internal models, and spaciousness gives rise to clarity, choosing not to act can be an intentional and powerful form of action. In a society that often prizes output, recognizing the value of pause (whether to reflect, recover, or realign) can be one of the most adaptive choices we make. Like every part of the active inference cycle, inaction plays a meaningful role in shaping how we learn, grow, and sustain ourselves.

**ANNOTATION**

That said, much has shifted in our broader environments while writing this book. Inaction has been important, but timely and thoughtful action has been critical as well.

## Know Your Strengths

As we explored in Chapter 25, our physiological systems vary—much like eye color or gait. Intelligence, too, follows this path. Drawing from Howard Gardner's Theory of Multiple Intelligences, we can appreciate that different people have different domains in which they thrive: linguistic, musical, spatial, kinesthetic, interpersonal, intrapersonal, naturalistic, and beyond. Each reflects a domain-specific generative model, detailing the ways our brains interpret and bodies sense and act upon the world. Recognizing this expands the idea of intelligence to include many forms that offer valuable strategies for navigating uncertainty and life. Statistically speaking, this also implies something hopeful: **that every person carries areas of strength**. Our opportunity is to discover them, nurture them, and share them.

**ANNOTATION**

From an active inference perspective, intelligence is the capacity to create effective models that minimize free energy in a particular context.

## Foster Community

As we've seen in Chapters 16 and 21, our nervous systems are inherently social. We co-regulate. We mirror. We learn in relation to others. From the way infants first attune to caregivers to the nuanced dance of collaboration in adulthood, our active inference cycles are built in tandem with those around us. Intentional communities act as both a buffer and a catalyst. They reduce uncertainty through shared meaning, reinforce adaptive models through feedback, and offer emotional scaffolding when we face challenges. Across religious traditions, psychological frameworks, and social structures, the value of human connection is clear. Taking the time to build environments and relationships that genuinely support your system can have a lasting impact. Emotional safety, perspective-shifting connection, and steady, grounded presence all can contribute to reducing tension and restoring balance in both isolated moments and across your life as a whole. These forms of support, both practical and relational, can be among the most powerful resources available to you.

## Closing

Being human is hard. No matter what your current call to action is, no matter the circumstances you were born into, the truth remains: suffering is not relative. We all suffer. Each of us is running a complex active inference cycle, shaped by past experience, environment, physiology, and belief. These cycles, beautiful and intricate, also bring with them pain, confusion, and dissonance. And yet, these same cycles hold within them the capacity for transformation. From a free energy perspective, pain often arises from prediction errors when what we expect of ourselves, others, or the world does not match what unfolds. The opportunity, then, is in the update. To meet those mismatches not with shame or resignation, but with curiosity, courage, and the willingness to evolve. As we move forward into a time of profound change, it feels to me that our greatest shared task is this: **to develop the capacity to stay with the challenging sensations these shifts bring.** To update. To adapt.

Life, in this sense, becomes an invitation. A chance to feel deeply alive in our bodies. To revel in the richness of sensation, to savor the warmth of sun on skin, the rhythm of laughter, the scent of something beloved. To reconnect with the vibrant aliveness that pulses through us when we're attuned. To find refuge in meaningful connection. To grow with intention. To create with meaning. To simply be human.

## SPARK

✦ Consider the content of this book. How might you want to engage with it going forward? Would it be helpful to set an intention to revisit something in a few months? Without being directive, how might you share what's been impactful for you with others?

THANK YOU
# KEEPING THIS BOOK ACCESSIBLE

Thank you for the purchase of this book. In an effort to keep it as accessible as possible, we priced the book at the cost of printing and distribution only (essentially selling it for less than it costs to make). If you would like to make a donation to help cover editorial, illustrative, research, and development costs or future research, you can do so at: **actincycle.com/donate**

# ACKNOWLEDGMENTS

I want to begin by expressing my deepest gratitude to my husband, William Carbone. Your unwavering dedication to finding coherence within yourself and between us, the gift of your deeply beautiful love, and your transformation into becoming a pivotal anchor amidst the uncertainty in my life are the most precious of gifts. I could not have done this without you. Thank you.

I want to acknowledge the many gifts my family has given me. I'm grateful to my parents for doing their absolute best in the ways they knew. My brother and his wife for making space for me as I am. My Portuguese family, who welcomed me into their home and taught me the beauty of simple family rituals. My American family, whose festive holiday tables brought so much warmth and delight and to my husband's family, for the fun and delicious times in the Italian kitchen.

In so many ways, this book would not have been possible without Karl Friston. I am grateful for the gift of your time and ideas to the world, and for shaping them in ways that remain accessible to many. Your dedication to using this work in service of others is, to me, your most inspiring quality.

I have immense gratitude to the remainder of the team behind this book, whose openness to doing things differently, dedication to deadlines, and commitment to the creative process made it possible. Ludo Cestarelli, thank you for your creativity, design talent, and the epic sprint you made in bringing this book to fruition. Zach Daugherty, thank you for working creatively with me and for your talent with words, which helped bring this book to life. Finally, I want to thank Sylvia Zhang, thank you for your friendship, ideas, and all the work you did on the references. The book would not be anywhere near the same without you three.

With deep gratitude to Yossi Vardi, who doesn't like compliments, so I'll keep this short: thank you. To the KN Community. This book would not exist without you, and neither would the person I've become. Thank you for giving me space to belong, connect, grow, and share ideas. You are a truly special group of people, and I hope to keep gathering with you often.

To the Ozyjowski family: your friendship, connection, and support through the years have been such a precious gift. Thank you for believing in me and supporting me even when my path diverged from the norm.

My deepest gratitude and acknowledgment to those listed below. You, your ideas, and our moments of connection in some way rippled into this work. Thank you.

Ada Parris, Agnese Sardella, Alberto-Giovanni Busetto, Alice Agogino, Alicia Navarro, Allan Davidson, Alex Haw, Andy Chen, Ana Meira, Andrea Mignolo, Andrew Wolfram, Angella Okawa, Anita Puppe, Assaf Biderman, Barak Berkowitz, Betinah Monteiro Pereira, Brendan DeWolf, Brett Christie, Carlo Rizzo, Chris Wray, Daniel Friendman, David Corr, David Grebler, David Rempel,

**Navigating Uncertainty**

Debu Purkayastha, Ellen Demlow, Eliram Haklai, Elnea Alessandra Vitale, Eloi Le Roux, Eric MacIntosh, Franziska Deecke, Fred Destin, Giulia Galbarini, Gil Golan, Gina Reimann, Gitte Fuglsang, Goliath Nadbornik, Gregorio Ameyugo, Guido Gruesshaber, Guido van Nispen, Heikki Haldre, Hedi Mardisoo, Hiroshi Ishii, Isaiah Fisher-Brown, Janma Bardi Aparicio, Jeff Connolly, Jeff Krone, Jennifer Antunes, Jeremy Slocum, Jernej Pintar, Jill Hellman, Joana Carolina Ribeiro Nunes da Silva de Gusmão, John Henry Clippinger, Jonas Mago, Jorge Epifanio, Jorge Oller, JP Rangaswami, Juca de Roos Schweitzer, Kamal Bherwani, Kevin Varend, Kristen Davis, Kunal Gupta, Laura Bechthold, Leanne Smullen, Limor Schweitzer, Louise Kahn, Louis Barinaga, Mari Jokiranta, Mark Clift, Marta Gajecka, Mary Glass, Matilda Lee, Megan Smith, Michael Abrash, Michael Pate, Michael Shiloh, Michelle de Bruijn, Mike Sparandara, Nadiia Kolibaba, Nathalie Kazzi, Nili Ohayon, Omri Shacham, Or Daniel, Patrick Philipp Eichiner, Raz Manasherov, Renuka Cavadini, Roberto Saint-Malo, Robbie Kimm, Robert Wolcott, Ross Palmer, Ryan Smith, Sahar Azarabadi, Salome Steinmann, Sam Glassenberg, Sanketh Shetty, Shannon Brown, Simon Stark, Sophia Swire, Sten Saluveer, Stephanie Hornung, Steven J. Posner, Susan Hasty, Suzanne Palmer, Sumit Jamuar, Tamas David-Barrett, Thomas Kehler, Tobias Burkhardt, Tomás Pereira da Silva, Vanessa Arelle, Vincent Favrat, Vishal Nangalia, Won Hee Chang, Yann Riche, Yaron Haklai, Yip Thy Diep Ta, Yossi Leshem, Yuki Gorospe.

# REFERENCES

**CHAPTER 1**

**To make things even more confusing, these demands are often presented amidst conflicting cues and signals.** Seth, Anil 'Interoceptive Inference, Emotion, and the Embodied Self'. *Trends in Cognitive Sciences* 17, no. 11 (2013): 565–73.

**Like many skilled musicians, this understanding can be embodied: supported both by mental cognition and bodily action and sensation.** Varela, Francisco, Evan Thompson, and Eleanor Rosch. *The Embodied Mind, Revised Edition: Cognitive Science and Human Experience.* MIT press, 2017.

**It anchors itself in active inference and free energy—two foundational principles in science that are often shrouded by advanced formulas and abstract applications.** Friston, Karl. 'The Free-Energy Principle: A Unified Brain Theory?' *Nature Reviews Neuroscience* 11, no. 2 (2010): 127–38; Friston, Karl, Thomas FitzGerald, Francesco Rigoli, Philipp Schwartenbeck, and Giovanni Pezzulo. 'Active Inference: A Process Theory'. *Neural Computation* 29, no. 1 (2017): 1–49.

**CHAPTER 2**

**It revealed how we navigate uncertainty, adapt to change, and have the opportunity to shift our lived reality.** Pezzulo, Giovanni, Thomas Parr, and Karl Friston. 'Active Inference as a Theory of Sentient Behavior'. *Biological Psychology* 186 (2024): 108741.

**CHAPTER 3**

**One is through foundational mathematics paired with the ability to test hypotheses, an approach on which active inference is based.** Buckley, Christopher, Chang Sub Kim, Simon McGregor, and Anil Seth. 'The Free Energy Principle for Action and Perception: A Mathematical Review'. *Journal of Mathematical Psychology* 81 (2017): 55–79.

**Active inference and free energy also achieve robust science by being supported by multiple peer-reviewed publications, with few contradictory findings, and reinforced by overwhelming additive results across diverse fields of study.** Friston, Karl. 'The Free-Energy Principle: A Unified Brain Theory?' *Nature Reviews*

*Neuroscience* 11, no. 2 (2010): 127–38; Hodson, Rowan, Marishka Mehta, and Ryan Smith. 'The Empirical Status of Predictive Coding and Active Inference'. *Neuroscience & Biobehavioral Reviews* 157 (2024): 105473.

**For example, a quick Google Scholar search reveals over 16,400 results for "Free Energy Principle," highlighting its substantial and rigorous area of research.** 'Free Energy Principle - Google Scholar'. Accessed 15 September 2025. https://scholar.google.com/scholar?hl=en&as_sdt=0%2C5&q=Free+Energy+Principle&oq=.

**The Free Energy Principle and Active Inference Cycle were first articulated and developed mathematically by neuroscientist Karl Friston primarily during the early 2000s.** Friston, Karl. 'A Theory of Cortical Responses'. *Philosophical Transactions of the Royal Society B: Biological Sciences* 360, no. 1456 (2005): 815–36.

**It typically takes between 40 and 100 years for new theoretical concepts to become part of the common vernacular, and at the time of writing, we're still relatively early.** Petersen, Alexander, Joel Tenenbaum, Shlomo Havlin, and Eugene Stanley. 'Statistical Laws Governing Fluctuations in Word Use from Word Birth to Word Death'. *Scientific Reports* 2, no. 1 (2012): 313; Rogers, Everett M. *Diffusion of Innovations, 5th Edition*. Free Press, 2003.

**Karl himself is a British neuroscientist trained in both medicine and mathematics, whose career has been defined by bridging rigorous theory with practical applications. Known for his meticulous methods and cross-disciplinary approach, he integrates insights from physics, neuroscience, and information theory to explain how living systems sustain themselves. He's often regarded as the "grandfather" of the Free Energy Principle and one of the most cited and influential scientists today.** Research.Com. 'Karl J. Friston: Neuroscience H-Index & Awards - Academic Profile'. Accessed 26 August 2025. https://research.com/u/karl-j-friston.

**Active Inference and Free Energy offer insightful models for understanding how living systems manage uncertainty. Fundamentally, these theories demonstrate that we, and other life, manage change and unknowns by aligning internal expectations (what we think will happen) with external reality (what actually happens). We then work to actively reduce discrepancy between the two.** Parr, Thomas, Giovanni Pezzulo, and Karl Friston. *Active Inference: The Free Energy Principle in Mind, Brain, and Behavior*. The MIT Press, 2022.

**Active Inference suggests that consciously holding various possible outcomes in mind, without prematurely settling on one, helps us remain balanced and responsive.** Friston, Karl, Lancelot Da Costa, Danijar Hafner, Casper Hesp, and Thomas Parr. 'Sophisticated Inference'. *Neural Computation* 33, no. 3 (2021): 713–63.

**Emotionally, this means being able to experience and process feelings like anxiety or confusion without immediately seeking resolution.** Hirsh, Jacob B., Raymond A. Mar, and Jordan B. Peterson. 'Psychological Entropy: A Framework for Understanding Uncertainty-Related Anxiety'. *Psychological Review* 119, no. 2 (2012): 304–20.

**References**

*In terms of Free Energy, maintaining open rather than fixating prematurely reduces internal prediction errors, essentially lowering our likelihood of incorrect predictions about the world. This lowers our free energy, the felt tension in our lives, and enhances our overall sense of stability and connection over time.* Millidge, Beren, Alexander Tschantz, and Christopher L. Buckley. 'Whence the Expected Free Energy?' *Neural Computation* 33, no. 2 (2021): 447–82.

[We also want to acknowledge that, given the realistic limitations of self-funded research required for the complex systems thinking of this book, we used AI as a supporting tool. Human authors and experts were always the originators, reviewers, and final approvers of the work.] ChatGPT, Gemini, Scite, and Claude were utilized for supplement support at various times between January 2023-September 2025.

### CHAPTER 4

**Embodying, however, goes beyond understanding to deeply integrate knowledge into daily actions and habits, aligning thought, behavior, and being, such as playing a deeply expressive and mastered piece.** Harrison, Paul. 'Making Sense: Embodiment and the Sensibilities of the Everyday'. *Environment and Planning D: Society and Space* 18, no. 4 (2000): 497–517.

**By consciously interacting with these materials, you nurture the integration of mind and body, fostering a deeper, more intuitive understanding of yourself and your environment.** Thompson, Evan. *Mind in Life: Biology, Phenomenology, and the Sciences of Mind.* Harvard University Press, 2010.

**In turn, these opportunities not only resonate intuitively but, as integrations of mind, body, and environment, are fundamentally supported by active inference.** Varela, Francisco, Evan Thompson, and Eleanor Rosch. *The Embodied Mind, Revised Edition: Cognitive Science and Human Experience.* MIT press, 2017.

### CHAPTER 5

**The active inference cycle is a scientific concept used to describe how our brains continuously sense and interpret the world around us, test those interpretations through our actions, and then adjust based on the feedback we receive from our environment, body, and mind.** Pezzulo, Giovanni, Thomas Parr, and Karl Friston. 'The Evolution of Brain Architectures for Predictive Coding and Active Inference'. *Philosophical Transactions of the Royal Society B* 377, no. 1844 (2021): 20200531.

**Much like the felt complexity of life, active inference spans multiple realms, including physics. What makes this particularly useful and fascinating is that the human brain is highly adept at making predictions and inferences rooted in physics.** Friston, Karl, Biswa Sengupta, and Gennaro Auletta. 'Cognitive Dynamics: From Attractors to Active Inference'. *Proceedings of the IEEE* 102, no. 4 (2014): 427–45.

**Navigating Uncertainty**

**The Mind: This is where we make sense of the world, form expectations about what might happen, and align our goals with our actions.** Clark, Andy. 'The Imaginarium'. *In Surfing Uncertainty: Prediction, Action, and the Embodied Mind.* Oxford University Press, 2015.

**Through action and sensation, it bridges the gap between the mind and environment, providing continuous feedback. From subtle interoceptive signals to the tangible effects of our movements, they shape our ongoing predictions and refine our responses.** Seth, Anil, and Manos Tsakiris. 'Being a Beast Machine: The Somatic Basis of Selfhood'. *Trends in Cognitive Sciences* 22, no. 11 (2018): 969–81.

**The Environment: External factors, including our relationships, culture, and surroundings, act as both the source of inputs and the context for our actions. The environment around us profoundly influences both our mind and body, shaping how we perceive, react, and adapt.** Vasil, Jared, Paul Badcock, Axel Constant, Karl Friston, and Maxwell Ramstead. 'A World Unto Itself: Human Communication as Active Inference'. *Frontiers in Psychology* 11 (2020): 417.

**One such example is the gut-brain axis: bodily states, such as digestion, can affect mental clarity, while mental clarity can directly impact gut health. Moreover, the environment provides the gut microbes essential to the gut-brain axis, further highlighting the dynamic interplay between these spaces.** Carabotti, Marilia, Annunziata Scirocco, Maria Antonietta Maselli, and Carola Severi. 'The Gut-Brain Axis: Interactions between Enteric Microbiota, Central and Enteric Nervous Systems'. *Annals of Gastroenterology: Quarterly Publication of the Hellenic Society of Gastroenterology* 28, no. 2 (2015): 203–9.

CHAPTER 6

**The free energy principle suggests that every action, thought, and feeling is part of a larger effort to reduce surprise and uncertainty in your life. Its function is to align your internal expectations with the world around you. When things don't go as expected, you feel discomfort or tension (free energy), prompting you to act, reflect, or adapt to bring yourself back into balance.** Friston, Karl. 'The Free-Energy Principle: A Unified Brain Theory?' *Nature Reviews Neuroscience* 11, no. 2 (2010): 127–38; Friston, Karl, Lancelot Da Costa, Noor Sajid, et al. 'The Free Energy Principle Made Simpler but Not Too Simple'. *Physics Reports*, 1024 (2023): 1-29.

**Instead of wanting more free energy in our lives, we generally seek less. We want to avoid excessively wasted energy, large unexpected surprises, or continued themes of dissonance—a persistent sensation that feels out of sync or carries risk to our overall stability.** Friston, Karl, Thomas FitzGerald, Francesco Rigoli, Philipp Schwartenbeck, John O' Doherty, and Giovanni Pezzulo. 'Active Inference and Learning'. *Neuroscience & Biobehavioral Reviews* 68 (2016): 862–79.

**References**

**However, some surprise plays a foundational role in how we make sense of the world and make decisions.** Clark, Andy. 'A Nice Surprise? Predictive Processing and the Active Pursuit of Novelty'. *Phenomenology and the Cognitive Sciences* 17, no. 3 (2018): 521–34.

**Overall, we make sense of things by finding explanations that render our observations the least uncertain while also acting in ways that maximize a specific kind of surprise.** Friston, Karl, Francesco Rigoli, Dimitri Ognibene, Christoph Mathys, Thomas Fitzgerald, and Giovanni Pezzulo. 'Active Inference and Epistemic Value'. *Cognitive Neuroscience* 6, no. 4 (2015): 187–214.

**Taking this concept a level deeper, the free energy principle suggests that the undercurrent of our lives isn't centered around optimizing for external markers like wealth or status, as society often emphasizes, but rather on reducing uncertainty and resolving internal instability.** Colombo, Matteo, and Cory Wright. 'First Principles in the Life Sciences: The Free-Energy Principle, Organicism, and Mechanism'. *Synthese* 198, no. 14 (2021): 3463–88.

**Engaging with the world for intrinsic value (such as seeking knowledge and understanding) resonates more with this principle than chasing external rewards, such as promotions or superficial beauty standards.** Ryan, Richard, and Edward Deci. 'Self-Determination Theory and the Facilitation of Intrinsic Motivation, Social Development, and Well-Being'. *The American Psychologist* 55, no. 1 (2000): 68–78.

**CHAPTER 7**

**One of those supportive processes is homeostasis, a quiet, behind-the-scenes process that keeps us alive and steady. It allows our bodies and minds to recalibrate continuously, helping us function smoothly even in the face of daily disruptions. Without it, balance would break down, leading to dysfunction and eventually the end of life.** Billman, George 'Homeostasis: The Underappreciated and Far Too Often Ignored Central Organizing Principle of Physiology'. *Frontiers in Physiology* 11 (2020): 200; Torday, John 'Homeostasis as the Mechanism of Evolution'. *Biology* 4, no. 3 (2015): 573–90.

**It ensures that the body, and the embodied brain, can adjust, reorganize, and recover when challenged. This adaptive ability, often referred to as allostasis, is what enables us to move through difficulty and return to center.** Katsumi, Yuta, Jordan Theriault, Karen Quigley, and Lisa Feldman Barrett. 'Allostasis as a Core Feature of Hierarchical Gradients in the Human Brain'. *Network Neuroscience* 6, no. 4 (2022): 1010–31; McEwen, Bruce S. 'Allostasis and Allostatic Load: Implications for Neuropsychopharmacology'. *Neuropsychopharmacology* 22, no. 2 (2000): 108–24.

**All life engages in homeostasis in some form.** Asarian, Gloy, and Geary Ramachandran. 'Homeostasis'. In *Encyclopedia of Human Behavior (Second Edition)*, edited by V. S. Ramachandran. Academic Press, 2012.

**Navigating Uncertainty**

**Our cognitive well-being depends on a delicate dance between predictability and novelty. Too much uncertainty can overwhelm us; too much sameness can stunt growth.** Peters, Achim, Bruce McEwen, and Karl Friston. 'Uncertainty and Stress: Why It Causes Diseases and How It Is Mastered by the Brain'. *Progress in Neurobiology* 156 (2017): 164–88; Schwartenbeck, Philipp, Johannes Passecker, Tobias Hauser, Thomas FitzGerald, Martin Kronbichler, and Karl Friston. 'Computational Mechanisms of Curiosity and Goal-Directed Exploration'. *eLife* 8 (2019): e41703.

**When ignored, they disrupt the broader cycle of active inference, limiting our potential.** Tschantz, Alexander, Laura Barca, Domenico Maisto, Christopher Buckley, Anil Seth, and Giovanni Pezzulo. 'Simulating Homeostatic, Allostatic and Goal-Directed Forms of Interoceptive Control Using Active Inference'. *Biological Psychology* 169 (2022): 108266.

**This includes essentials such as water free of toxins or harmful pathogens, as well as physical spaces that protect us from violence, abuse, or chronic fear.** Hutton, Guy, and Claire Chase. 'Water Supply, Sanitation, and Hygiene'. In *Injury Prevention and Environmental Health*, 3rd ed., edited by Charles N. Mock, Rachel Nugent, Olive Kobusingye, and Kirk R. Smith. The International Bank for Reconstruction and Development / The World Bank, 2017; Shonkoff, Jack, Andrew Garner, Committee on Psychosocial Aspects of Child and Family Health, Committee on Early Childhood, Adoption, and Dependent Care, and Section on Developmental and Behavioral Pediatrics. 'The Lifelong Effects of Early Childhood Adversity and Toxic Stress'. *Pediatrics* 129, no. 1 (2012): e232-246.

**Our environments also need to meet our psychological needs, providing social bonds rooted in trust, a sense of belonging within a group or culture, and mutual support in times of need.** Deci, Edward, and Richard Ryan. 'Self-Determination Theory'. In *Handbook of Theories of Social Psychology* 1, no. 20 (2012): 416-436.

**We need emotional safety and the ability to express ourselves without fear of harm or rejection. Stress regulation is also critical, whether through mental or physical coping mechanisms or supportive environments.** Leary, Mark 'Affiliation, Acceptance, and Belonging: The Pursuit of Interpersonal Connection'. In *Handbook of Social Psychology, Vol. 2, 5th Ed.* John Wiley & Sons, Inc., 2010.

**For many, emotional needs extend to a sense of purpose, meaning, or inner harmony.** Fang, Louis, Alfred Allan, and Joanne Dickson. 'Purpose in Life and Associated Cognitive and Affective Mechanisms'. *Journal of Happiness Studies* 25, no. 6 (2024): 63.

CHAPTER 8

**More specifically, self-knowing is an ongoing process of making sense of how your inner world and the outer world interact.** Schwengerer, Lukas. 'Self-Knowledge in a Predictive Processing Framework'. *Review of Philosophy and Psychology* 10, no. 3 (2019): 563–85.

**Grounded in the science of active inference, self-knowing means constantly updating your understanding based on new information. This could include your best guesses about yourself and the world or your**

**community.** Nesi, Jacson, Roberta Lemos dos Santos, and Michele Benites. 'Exploring Enactivism: A Scoping Review of Its Key Concepts and Theorical Approach'. *Advances in Integrative Medicine* 11, no. 4 (2024): 184–90; Ramstead, Maxwell, Michael Kirchhoff, and Karl Friston. 'A Tale of Two Densities: Active Inference Is Enactive Inference'. *Adaptive Behavior* 28, no. 4 (2020): 225–39.

**However, self-knowing is also greatly supported by enactivism: an interdependent interaction of mind, body, and environment.** Maiese, Michelle. 'Can the Mind Be Embodied, Enactive, Affective, and Extended?' *Phenomenology and the Cognitive Sciences* 17, no. 2 (2018): 343–61.

**These actions are opportunities for meaning-making and processes through which we learn, understand, and shape both ourselves and our environment. Through these interactions, self-knowing becomes an iterative process and it evolves as we engage with the world and refine our understanding of it.** Bruineberg, Jelle, Julian Kiverstein, and Erik Rietveld. 'The Anticipating Brain Is Not a Scientist: The Free-Energy Principle from an Ecological-Enactive Perspective'. *Synthese* 195, no. 6 (2018): 2417–44.

**At the most basic level, we can remember these processes involve the body sensing what's happening in the world through inputs like sight, sound, and interoceptive signals, and the mind supporting the creation of meaning from these sensations.** Seth, Anil, Keisuke Suzuki, and Hugo Critchley. 'An Interoceptive Predictive Coding Model of Conscious Presence'. *Frontiers in Psychology* 2 (2011): 395.

**One of the key filters we have the opportunity to shift through self-knowing is called precision—the level of confidence and focus we give our senses and interpretations.** Parr, Thomas, and Karl Friston. 'The Anatomy of Inference: Generative Models and Brain Structure'. *Frontiers in Computational Neuroscience* 12 (2018): 90.

**Self-knowing allows you to recognize and adjust the focus and confidence you assign to your perceptions.** Picard, Fabienne, and Karl Friston. 'Predictions, Perception, and a Sense of Self'. *Neurology* 83, no. 12 (2014): 1112–18.

**Studies demonstrate that just eight weeks of awareness-based meditation can produce measurable changes in the brain. [A notable study led by Hölzel and colleagues found that participants experienced increased gray matter density in the hippocampus, a region critical for learning and memory, and reduced gray matter in the amygdala, which plays a key role in stress and anxiety.]** Hölzel, Britta, James Carmody, Mark Vangel, et al. 'Mindfulness Practice Leads to Increases in Regional Brain Gray Matter Density'. *Psychiatry Research: Neuroimaging* 191, no. 1 (2011): 36–43; Raugh, Ian, Alysia Berglund, and Gregory Strauss. 'Implementation of Mindfulness-Based Emotion Regulation Strategies: A Systematic Review and Meta-Analysis'. *Affective Science* 6, no. 1 (2025): 171–200.

**[Beyond these measurable shifts, practitioners reported significant improvements in their experiential quality of life, including enhanced emotional regulation, mental clarity, and a greater sense of calm.]**

**Navigating Uncertainty**

Lykins, Emily, and Ruth Baer. 'Psychological Functioning in a Sample of Long-Term Practitioners of Mindfulness Meditation'. *Journal of Cognitive Psychotherapy* 23, no. 3 (2009): 226–41.

**On this view, meditation and other practices enable one to develop the skill of relaxing the confidence placed in prior interpretations by tuning back into our senses.** Manjaly, Zina-Mary, and Sandra Iglesias. 'A Computational Theory of Mindfulness Based Cognitive Therapy from the "Bayesian Brain" Perspective'. *Frontiers in Psychiatry* 11 (2020): 404.

**Meditation and similar activities also create a vital connection between our bodily sensations, mental interpretations, and the precision we assign to these experiences.** Deane, George, Mark Miller, and Sam Wilkinson. 'Losing Ourselves: Active Inference, Depersonalization, and Meditation'. *Frontiers in Psychology* 11 (2020): 539726.

**This heightened awareness recalibrates the confidence placed in entrenched beliefs or past assumptions that are often lurking and influencing our cycles below the surface.** Lutz, Antoine, Jérémie Mattout, and Giuseppe Pagnoni. 'The Epistemic and Pragmatic Value of Non-Action: A Predictive Coding Perspective on Meditation'. *Current Opinion in Psychology* 28 (2019): 166-171.


CHAPTER 9

**We can play with both the zoom (attention) and the focus (precision) of our intentional awareness, and by doing so, we gain tools to more skillfully understand and navigate our world. These ideas, drawn from the science of active inference, show that perception is both receptive and actively shaped by us.** Feldman, Harriet, and Karl Friston. 'Attention, Uncertainty, and Free-Energy'. *Frontiers in Human Neuroscience* 4 (2010): 215.

**Once you've zoomed in, the next step is to adjust the focus and precision. The sharper the focus, the more clarity we bring to what we're attending to—allowing us to more precisely understand and engage with what we've zoomed in and turned our attention to.** Mirza, Berk, Rick Adams, Karl Friston, and Thomas Parr. 'Introducing a Bayesian Model of Selective Attention Based on Active Inference'. *Scientific Reports* 9, no. 1 (2019): 13915.

**On a neurological level, precision fine-tunes the brain's signal strength, akin to the conductor directing an orchestra—ensuring key signals are amplified while irrelevant ones are dampened. Known as "synaptic gain," this process dynamically balances neural excitation and inhibition, prioritizing the most relevant information for action and perception.** Brown, Harriet, and Karl Friston. 'Dynamic Causal Modelling of Precision and Synaptic Gain in Visual Perception — an EEG Study'. *NeuroImage* 63, no. 1 (2012): 223–31.

**For example, consider saccadic suppression. As you're reading this book or moving through your day, your brain automatically blurs visual input each time your eyes shift.** Krekelberg, Bart. 'Saccadic Suppression'. *Current Biology* 20, no. 5 (2010): R228–29.

**References**

**Without this, the constant motion from your eye movements would likely overwhelm your visual system. This built-in mechanism prevents disorientation, allowing you to perceive a stable world even while your gaze is constantly in motion.** Wurtz, Robert, Wilsaan Joiner, and Rebecca Berman. 'Neuronal Mechanisms for Visual Stability: Progress and Problems'. *Philosophical Transactions of the Royal Society B: Biological Sciences* 366, no. 1564 (2011): 492–503.

**Techniques such as mindfulness and meditation act like workouts for our minds, helping us deliberately fine-tune attention and clarity.** Holas, Paweł, and Justyna Kamińska. 'Mindfulness Meditation and Psychedelics: Potential Synergies and Commonalities'. *Pharmacological Reports* 75, no. 6 (2023): 1398–409.

**As highlighted in recent research, mindfulness meditation can improve your capacity to refine these predictions, making your mental models more accurate and adaptable. Over time, this repeated updating strengthens your ability to maintain clarity and helps you avoid becoming stuck in outdated or unhelpful patterns of thought.** Lutz, Antoine, Oussama Abdoun, Yair Dor-Ziderman, Fynn-Mathis Trautwein, and Aviva Berkovich-Ohana. 'An Overview of Neurophenomenological Approaches to Meditation and Their Relevance to Clinical Research'. *Biological Psychiatry: Cognitive Neuroscience and Neuroimaging*, Cognitive Neuroscience of Mindfulness, 10, no. 4 (2025): 411–24; Lutz, Antoine, Jérémie Mattout, and Giuseppe Pagnoni. 'The Epistemic and Pragmatic Value of Non-Action: A Predictive Coding Perspective on Meditation'. *Current Opinion in Psychology* 28 (2019): 166-171.

**Zoom and focus don't just shape personal perception; they extend outward, influencing collective and cultural narratives. On a societal level, shared attention guides what issues, values, or stories become priorities.** Shteynberg, Garriy. 'A Collective Perspective: Shared Attention and the Mind'. *Current Opinion in Psychology* 23 (2018): 93-7. Bouizegarene, Nabil, Maxwell Ramstead, Axel Constant, Karl Friston, and Laurence Kirmayer. 'Narrative as Active Inference: An Integrative Account of Cognitive and Social Functions in Adaptation'. *Frontiers in Psychology* 15 (2024): 1345480..

**Conversely, when attention becomes scattered or misaligned (as seen with sensationalized media or fear-driven narratives) it can lead to increased uncertainty, confusion, and anxiety.** Dabas, Pooja. 'A Brief Study on Whether Media Enhance or Prevent Panic During Disasters'. *Integrated Journal for Research in Arts and Humanities* 2, no. 4 (2022): 168–74; Kesner, Ladislav, Veronika Juríčková, Dominika Grygarová, and Jiří Horáček. 'Impact of Media-Induced Uncertainty on Mental Health: Narrative-Based Perspective'. *JMIR Mental Health* 12 (2025): e68640.

**Honing the ability to prioritize (zoom) and clarify (focus) signals from both internal and external environments aligns perception more closely with reality, reduces uncertainty, and strengthens adaptive responses to life's changes.** Sandved-Smith, Lars, Casper Hesp, Jérémie Mattout, Karl Friston, Antoine Lutz, and Maxwell Ramstead. 'Towards a Computational Phenomenology of Mental Action: Modelling Meta-Awareness and Attentional Control with Deep Parametric Active Inference'. *Neuroscience of Consciousness* 2021, no. 1 (2021): niab018.

That cell has its own boundaries and interacts with the world around it (like absorbing nutrients and expelling waste) while maintaining its internal stability. It has a kind of "skin" or layer separating it from the rest of the body, defining what is "inside" the cell and what is "outside." This boundary is its Markov blanket, enabling it to interact with its environment while preserving its own integrity. Friston, Karl. 'Life as We Know It'. *Journal of The Royal Society Interface* 10, no. 86 (2013) 20130475.

In this view, your Markov blanket becomes the boundary between you and your environment, helping you interpret, respond to, and regulate your relationship with the world around you. Palacios, Ensor Rafael, Adeel Razi, Thomas Parr, Michael Kirchhoff, and Karl Friston. 'On Markov Blankets and Hierarchical Self-Organisation'. *Journal of Theoretical Biology* 486 (2020): 110089.

As humans, collectively, we are part of the Earth's ecosystems, interconnected through shared resources, climate, and social structures. From this perspective, our species as a whole can be seen as having a kind of Markov blanket, where interactions occur at the boundary between humanity and everything else. Rubin, Sergio, Thomas Parr, Lancelot Da Costa, and Karl Friston. 'Future Climates: Markov Blankets and Active Inference in the Biosphere'. *Journal of The Royal Society Interface* 17, no. 172 (2020): 20200503.

Scientifically speaking, all three perspectives are real and equally valued: my cell in my body, me as an individual, and me as a member of the human species. Palacios, Ensor Rafael, Adeel Razi, Thomas Parr, Michael Kirchhoff, and Karl Friston. 'On Markov Blankets and Hierarchical Self-Organisation'. *Journal of Theoretical Biology* 486 (2020): 110089.

Thinking about Markov blankets at different scales (and the idea of Markov blankets within Markov blankets) can illustrate the seamless, interconnected way we self-organize at cellular, personal, and societal levels. Ramstead, Maxwell, Paul Badcock, and Karl Friston. 'Answering Schrödinger's Question: A Free-Energy Formulation'. *Physics of Life Reviews* 24 (2018): 1–16.

They enable interaction, adaptation, and resilience, allowing systems to persist even amidst constant change. This dual role of boundaries, as both protectors of internal integrity and facilitators of external connections, underscores their dynamic nature and the need for balance between the external and internal. Kirchhoff, Michael, Thomas Parr, Ensor Palacios, Karl Friston, and Julian Kiverstein. 'The Markov Blankets of Life: Autonomy, Active Inference and the Free Energy Principle'. *Journal of The Royal Society Interface* 15, no. 138 (2018): 20170792.

The systems and cultural norms around us shape our behavior and guide what's considered appropriate, possible, or even visible to us (top-down causation). At the same time, our individual actions ripple outward, co-constructing and reinforcing those very structures (bottom-up causation). Kaufmann, Rafael, Pranav Gupta, and Jacob Taylor. 'An Active Inference Model of Collective Intelligence'. *Entropy* 23, no. 7 (2021): 830.

**References**

**Opportunities to resolve uncertainties can be divided into two broad categories: utilization and exploration. Utilization involves using existing knowledge to navigate familiar situations effectively. For example, consider cooking a meal you've made many times before. You know the recipe, the steps, and the expected outcomes, making the process predictable and stable. Each time you cook, you reinforce your ability to handle this specific situation, building confidence and efficiency over time. Exploration, on the other hand, involves seeking out new experiences or information to expand your understanding of the world.** Mehlhorn, Katja, Ben Newell, Peter Todd, et al. 'Unpacking the Exploration–Exploitation Tradeoff: A Synthesis of Human and Animal Literatures.' *Decision* 2, no. 3 (2015): 191–215.

**Balancing familiarity with novelty is key for responding effectively to uncertainty; without this balance, we risk stagnation in outdated cycles or becoming overwhelmed by constant change.** Jami, Ali, Sajjad Abbaszade, and Abdol-Hossein Vahabie. 'A Review on Exploration–Exploitation Trade-off in Psychiatric Disorders'. *BMC Psychiatry* 25, no. 1 (2025): 420.

**For example, repeated exposure to uncertainty or feared situations activates the prefrontal cortex, which in turn regulates the amygdala. This process aids fear extinction and helps new experiences feel less threatening over time.** Kenwood, Margaux, Ned Kalin, and Helen Barbas. 'The Prefrontal Cortex, Pathological Anxiety, and Anxiety Disorders'. *Neuropsychopharmacology* 47, no. 1 (2022): 260–75; Landowska, Aleksandra, David Roberts, Peter Eachus, and Alan Barrett. 'Within- and Between-Session Prefrontal Cortex Response to Virtual Reality Exposure Therapy for Acrophobia'. *Frontiers in Human Neuroscience* 12 (2018): 362; Milad, Mohammed, and Gregory Quirk. 'Fear Extinction as a Model for Translational Neuroscience: Ten Years of Progress'. *Annual Review of Psychology* 63, no. 1 (2012): 129–51; Morriss, Jayne. 'Psychological Mechanisms Underpinning Change in Intolerance of Uncertainty across Anxiety-Related Disorders: New Insights for Translational Research'. *Neuroscience & Biobehavioral Reviews* 173 (2025): 106138; Ueda, Kazutaka, Takahiro Sekoguchi, and Hideyoshi Yanagisawa. 'How Predictability Affects Habituation to Novelty'. *PLOS ONE* 16, no. 6 (2021): e0237278.

**Additionally, habituation (the brain's ability to normalize new experiences) reduces the discomfort of novelty, fostering resilience and adaptability.** Ueda, Kazutaka, Takahiro Sekoguchi, and Hideyoshi Yanagisawa. 'How Predictability Affects Habituation to Novelty'. *PLOS ONE* 16, no. 6 (2021): e0237278.

**Engaging with uncertainty can also activate the brain's reward system, releasing dopamine and reinforcing the positive effects of exploration and adaptation. Over time, these neural processes enhance emotional stability, improve specific skills, and equip the brain to respond more effectively to unexpected challenges.** DeYoung, Colin 'The Neuromodulator of Exploration: A Unifying Theory of the Role of Dopamine in Personality'. *Frontiers in Human Neuroscience* 7 (2013): 762; Schwartenbeck, Philipp, Johannes Passecker, Tobias U Hauser, Thomas FitzGerald, Martin Kronbichler, and Karl Friston. 'Computational Mechanisms of Curiosity and Goal-Directed Exploration'. *eLife* 8 (2019): e41703; Wang, Yuhao, Armin Lak, Sanjay Manohar, and Rafal Bogacz. 'Dopamine Encoding of Novelty Facilitates Efficient Uncertainty-Driven Exploration'. *PLOS Computational Biology* 20, no. 4 (2024): e1011516.

**However, building these habits and inciting shifts requires both time and effort, as they involve physical changes in the brain.** Mendelsohn, Alana. 'Creatures of Habit: The Neuroscience of Habit and Purposeful Behavior'. *Biological Psychiatry* 85, no. 11 (2019): e49–51.

**Research consistently shows that successful habit formation begins with incremental steps.** Gardner, Benjamin, Phillippa Lally, and Jane Wardle. 'Making Health Habitual: The Psychology of "Habit-Formation" and General Practice'. Debate & Analysis. *British Journal of General Practice* 62, no. 605 (2012): 664–66.

**Our brains are constantly making predictions about the world around us based on prior experiences and beliefs.** Sprevak, Mark, and Ryan Smith. 'An Introduction to Predictive Processing Models of Perception and Decision-Making'. *Topics in Cognitive Science* (2023).

**Without these updates, we risk getting stuck in cycles of outdated information that fail to resolve the uncertainties we encounter.** Kube, Tobias, and Liron Rozenkrantz. 'When Beliefs Face Reality: An Integrative Review of Belief Updating in Mental Health and Illness'. *Perspectives on Psychological Science* 16, no. 2 (2021): 247–74.

**CHAPTER 12**

**A key component of this process is empathy. Often thought to be supported by mirror neurons, empathy allows us to simulate and predict the actions and emotions of others.** Bonini, Luca, Cristina Rotunno, Edoardo Arcuri, and Vittorio Gallese. 'Mirror Neurons 30 Years Later: Implications and Applications'. *Trends in Cognitive Sciences* 26, no. 9 (2022): 767–81.

**A favorite discovery story of mine, mirror neurons were first discovered in the 1990s by researchers studying monkeys. It was first inadvertently observed when a monkey's brain unexpectedly responded to an action performed by a researcher. The monkey was already hooked up to monitor its brain activity for a set of planned tests. However, when the researcher picked up an object and the monkey remained still (a condition not part of the initially planned testing sequence), the same neural pattern fired as if the monkey itself had performed the action. This led to the inadvertent discovery of mirror neurons.** Figueiredo, Luiz Felipe, Maria Eduarda Lannes, Catia Mathias, Marlede Mota Gomes, and Antonio Nardi. 'The mirror neuron: thirty years since its discovery'. *Brazilian Journal of Psychiatry* 45, no. 3 (2023): 298–99; Di Pellegrino, Giuseppe, Luciano Fadiga, Leonardo Fogassi, Vittorio Gallese, and Giacomo Rizzolatti. 'Understanding Motor Events: A Neurophysiological Study'. *Experimental Brain Research* 91, no. 1 (1992): 176–80.

**References**

These specialized brain cells activate both when an individual performs an action and when they observe someone else performing the same action. In other words, parts of our brains react the same whether we're experiencing something directly ourselves or empathizing with someone else. Cook, Richard, Geoffrey Bird, Caroline Catmur, Clare Press, and Cecilia Heyes. 'Mirror Neurons: From Origin to Function'. *Behavioral and Brain Sciences* 37, no. 2 (2014): 177–92.

One key concept often helpful to understand is that unpredicted discomfort is often more distressing than expected discomfort because the unexpected in itself is often uncomfortable. Berker, Archy de, Robb Rutledge, Christoph Mathys, et al. 'Computations of Uncertainty Mediate Acute Stress Responses in Humans'. *Nature Communications* 7, no. 1 (2016): 10996.

There's a second concept that often contributes to escalation: the brain filtering out sensory inputs during action, a process known as sensory attenuation. To act effectively, we often temporarily ignore the immediate sensory feedback of our actions. Roussel, Cedric, Gethin Hughes, and Florian Waszak. 'A Preactivation Account of Sensory Attenuation'. *Neuropsychologia* 51, no. 5 (2013): 922–29.

Sensory attenuation helps prioritize certain signals or intentions over others, enabling smooth initiation and execution of actions. In extreme cases, we can see how this process can break down, like Parkinson's disease, where the brain's inability to suppress sensory feedback can contribute to difficulty initiating movement. Kearney, Joshua, and John-Stuart Brittain. 'Sensory Attenuation in Sport and Rehabilitation: Perspective from Research in Parkinson's Disease'. *Brain Sciences* 11, no. 5 (2021): 580; Wolpe, Noham, Jiaxiang Zhang, Cristina Nombela, James Ingram, Daniel Wolpert, and James Rowe. 'Sensory Attenuation in Parkinson's Disease Is Related to Disease Severity and Dopamine Dose'. *Scientific Reports* 8, no. 1 (2018): 15643.

At its extreme, this dynamic leads to dehumanization, where the other person is no longer seen as fully human because their behavior seems fundamentally different from our own. This can prevent constructive dialogue and deepen divisions. When this happens, it poses significant risks by escalating conflicts, fostering alienation, and breaking down social cohesion. Haslam, Nick. 'Dehumanization and the Lack of Social Connection'. *Current Opinion in Psychology* 43 (2022): 312–16.

Recent scientific research describes this as attentional narrowing, a natural brain response driven by stress and uncertainty. Attentional narrowing helps explain why, under pressure, we often default to habitual patterns, even if they aren't the best choices available to us. Prinet, Julie, and Nadine Sarter. 'The Effects of High Stress on Attention: A First Step Toward Triggering Attentional Narrowing in Controlled Environments'. *Proceedings of the Human Factors and Ergonomics Society Annual Meeting* 59, no. 1 (2015): 1530–34.

When uncertainty spikes, our brains release chemicals like adrenaline and noradrenaline. These neurochemicals disrupt the normal functioning of the prefrontal cortex, the region associated with

**thoughtful decision-making and flexibility.** Starcke, Katrin, and Matthias Brand. 'Effects of Stress on Decisions under Uncertainty: A Meta-Analysis'. *Psychological Bulletin* (US) 142, no. 9 (2016): 909–33.

**Simultaneously, "neural gain" (the amplification of strong signals and suppression of weaker ones) causes familiar ideas and actions to dominate, effectively sidelining alternative perspectives.** Eldar, Eran, Yael Niv, and Jonathan Cohen. 'Do You See the Forest or the Tree? Neural Gain and Breadth Versus Focus in Perceptual Processing'. *Psychological Science* 27, no. 12 (2016): 1632–43.

**Research demonstrates that relationship uncertainty or financial instability can intensify our ideological convictions. When faced with uncertainty, our brain seeks comfort in certainty—even if that certainty involves polarized or extreme beliefs.** Hogg, Michael 'From Uncertainty to Extremism: Social Categorization and Identity Processes'. *Current Directions in Psychological Science* 23, no. 5 (2014): 338–42; Hogg, Michael, Arie Kruglanski, and Kees Van Den Bos. 'Uncertainty and the Roots of Extremism'. *Journal of Social Issues* 69, no. 3 (2013): 407–18.

**Intentionally zooming out to consider shared beliefs counteracts automatic narrowing, re-engages higher-order reasoning, and encourages neural flexibility. This cognitive shift decreases stress-induced rigidity, promoting enhanced creativity, improved problem-solving capabilities, and greater openness to collaborative interactions.** Vasil, Jared, Paul Badcock, Axel Constant, Karl Friston, and Maxwell Ramstead. 'A World Unto Itself: Human Communication as Active Inference'. *Frontiers in Psychology* 11 (2020): 417.

**The mirror neurons we described earlier provide us with the wiring to use empathy for our own benefit.** Iacoboni, Marco. 'Imitation, Empathy, and Mirror Neurons'. *Annual Review of Psychology* 60 (2009): 653–70.

**The free energy principle reminds us that our foundational primary drive is not external validation but coherence and reduced uncertainty. We can also combine an empathic lens with our understanding of how people's desire to reduce uncertainty is shaped by their past experiences and personal circumstances.** Holmes, Jeremy, and Tobias Nolte. '"Surprise" and the Bayesian Brain: Implications for Psychotherapy Theory and Practice'. *Frontiers in Psychology* 10 (2019): 592.

**CHAPTER 13**

**Contrary to the dated notion that our brain passively responds to the environment (a one-way transfer of information), modern neuroscience reveals a bidirectional process: the brain not only receives information from the environment but also actively projects expectations outward.** Bubic, Andreja, Yves Von Cramon, and Ricarda Schubotz. 'Prediction, Cognition and the Brain'. Frontiers in Human Neuroscience 4 (2010): 1094; Clark, Andy. 'Whatever next? Predictive Brains, Situated Agents, and the Future of Cognitive Science'. *Behavioral and Brain Sciences* 36, no. 3 (2013): 181–04.

**References**

**If we remember the cycle that is active inference, then one arrow at all times is pointing outward (what we expect) and another is pointing inwards (what we perceive and experience), helping to bring coherence and reduce the surprise as we navigate the world.** Da Costa, Lancelot, Thomas Parr, Noor Sajid, Sebastijan Veselic, Victorita Neacsu, and Karl Friston. 'Active Inference on Discrete State-Spaces: A Synthesis'. *Journal of Mathematical Psychology* 99 (2020): 102447.

**Our expectations aren't just predictions; they are active forces that shape what we see, hear, and experience. In a very real way, we author our perceptions by acting on the world to bring our expectations to life. And in turn, the world pushes back, refining and reshaping those expectations.** Bruineberg, Jelle. 'Active Inference and the Primacy of the 'I Can'?' *In Philosophy and Predictive Processing*, edited by Thomas Metzinger and Wanja Wiese. 2017; Friston, Karl, Philipp Schwartenbeck, Thomas Fitzgerald, Michael Moutoussis, Tim Behrens, and Raymond Dolan. 'The Anatomy of Choice: Active Inference and Agency'. *Frontiers in Human Neuroscience* 7 (2013): 598.

**Scientists often describe intentions as "priors over policies," meaning they are pre-existing beliefs that guide possible actions. These existing beliefs are shaped by our identity and anticipated outcomes, which explains why intentions often feel deeply personal and morph over time.** Friston, Karl, Philipp Schwartenbeck, Thomas FitzGerald, Michael Moutoussis, Timothy Behrens, and Raymond Dolan. 'The Anatomy of Choice: Dopamine and Decision-Making'. *Philosophical Transactions of the Royal Society B: Biological Sciences* 369, no. 1655 (2014): 20130481; Moutoussis, Michael, Pasco Fearon, Wael El-Deredy, Raymond Dolan, and Karl Friston. 'Bayesian Inferences about the Self (and Others): A Review'. *Consciousness and Cognition* 25 (2014): 67–76.

**This self-model—the narrative of who you are—acts as the foundation for your intentions, influencing not just what you do but how you perceive the outcomes of your actions.** Deane, George. 'Consciousness in Active Inference: Deep Self-Models, Other Minds, and the Challenge of Psychedelic-Induced Ego-Dissolution'. *Neuroscience of Consciousness* 2021, no. 2 (2021): niab024; Hohwy, Jakob. 'Conscious Self-Evidencing'. *Review of Philosophy and Psychology* 13, no. 4 (2022): 809–28.

**Intentions aligned with reducing system dissonance (rather than shallower societal ones) can better serve our systems at a fundamental level by resolving internal and external tension more effectively.** Peters, Achim, Bruce McEwen, and Karl Friston. 'Uncertainty and Stress: Why It Causes Diseases and How It Is Mastered by the Brain'. *Progress in Neurobiology* 156 (2017): 164–88; Smith, Ryan, Maxwell Ramstead, and Alex Kiefer. 'Active Inference Models Do Not Contradict Folk Psychology'. *Synthese* 200, no. 2 (2022): 81.

**These imperatives align with two types of intentions: pragmatic intentions (based on what is already known) and epistemic intentions (focused on seeking new knowledge). Pragmatic intentions aim to maximize expected outcomes using established information.** Friston, Karl 'Active Inference and Cognitive Consistency'. *Psychological Inquiry* 29, no. 2 (2018): 67–73.

**Intentions are not static; they are part of a dynamic feedback loop. As we act on our intentions, we refine them based on the outcomes we experience.** Friston, Karl, Tommaso Salvatori, Takuya Isomura, et al. 'Active Inference and Intentional Behavior'. *Neural Computation* 37, no. 4 (2025): 666–700.

**CHAPTER 14**

**Willpower, as it's traditionally understood, resides in the brain.** Posner, Michael, and Mary Rothbart. 'Willpower and Brain Networks'. *ISSBD Bulletin* 2012, no. 1 (2012): 7–10.

**The brain is incredibly complex, with distinct regions responsible for different functions. Notably, only about one-third of our brain mass (the prefrontal cortex) underpins willpower.** Donahue, Chad, Matthew Glasser, Todd Preuss, James Rilling, and David Van Essen. 'Quantitative Assessment of Prefrontal Cortex in Humans Relative to Nonhuman Primates'. *Proceedings of the National Academy of Sciences* 115, no. 22 (2018): E5183–92; Stuss, Donald, and Robert Knight, eds. *Principles of frontal lobe function.* Oxford University Press, 2013.

**The prefrontal cortex is performing multiple key tasks at any given moment, from problem-solving and decision-making to regulating emotions and managing social interactions. Willpower is just one of many functions it's juggling.** Heatherton, Todd, and Dylan Wagner. 'Cognitive Neuroscience of Self-Regulation Failure'. *Trends in Cognitive Sciences* 15, no. 3 (2011): 132–39; Miller, Earl, and Jonathan Cohen. 'An Integrative Theory of Prefrontal Cortex Function'. *Annual Review of Neuroscience* 24, no. 1 (2001): 167–202.

**For example, the limbic system, which is responsible for our immediate responses and emotional reactions, operates largely outside of conscious control. In other words, we largely can't control those immediate emotional reactions we feel. When we're tired, stressed, or overwhelmed, the reactive systems dominate, reducing the influence of the prefrontal cortex.** Arnsten, Amy 'Stress Signalling Pathways That Impair Prefrontal Cortex Structure and Function'. *Nature Reviews. Neuroscience* 10, no. 6 (2009): 410–22; Rajmohan, V., and E. Mohandas. 'The Limbic System'. *Indian Journal of Psychiatry* 49, no. 2 (2007): 132–39.

**Instead of over-relying on willpower, we can shift to volition—the ability to align our actions with our beliefs and goals.** Haggard, Patrick. 'Human Volition: Towards a Neuroscience of Will'. *Nature Reviews Neuroscience* 9, no. 12 (2008): 934–46.

**Firstly, it triggers the body's stress response, activating the sympathetic nervous system and elevating cortisol and adrenaline. Chronic activation can lead to fatigue, cognitive impairments, and compromised immune function.** James, Katharine Ann, Juliet Ilena Stromin, Nina Steenkamp, and Marc Irwin Combrinck. 'Understanding the Relationships between Physiological and Psychosocial Stress, Cortisol and Cognition'. *Frontiers in Endocrinology* 14 (2023): 1085950; Mariotti, Agnese. 'The Effects of Chronic Stress on Health: New Insights into the Molecular Mechanisms of Brain–Body Communication'. *Future Science OA* 1, no. 3 (2015): FSO23.

**Secondly, persistent misalignments result in prediction errors within our brain, demanding considerable physiological energy to resolve these discrepancies, leading to exhaustion.** Arnaldo, Irene, Andrew Corcoran, Karl Friston, and Maxwell Ramstead. 'Stress and Its Sequelae: An Active Inference Account of the Etiological Pathway from Allostatic Overload to Depression'. *Neuroscience and Biobehavioral Reviews* 135 (2022): 104590; Robinson, Oliver, Cassie Overstreet, Danielle Charney, Katherine Vytal, and Christian Grillon. 'Stress Increases Aversive Prediction Error Signal in the Ventral Striatum'. *Proceedings of the National Academy of Sciences* 110, no. 10 (2013): 4129–33.

**Thirdly, ongoing stress responses disrupt key brain regions, impairing emotional regulation and cognitive clarity, which perpetuates cycles of ineffective behavior.** Girotti, Milena, Sarah Bulin, and Flavia Carreno. 'Effects of Chronic Stress on Cognitive Function – From Neurobiology to Intervention'. *Neurobiology of Stress* 33 (2024): 100670; Ragen, B. J., A. E. Roach, and C. L. Chollak. 'Chapter 29 - Chronic Stress, Regulation of Emotion, and Functional Activity of the Brain'. In *Stress: Concepts, Cognition, Emotion, and Behavior*, edited by George Fink. Academic Press, 2016.

**Finally, misalignments disturb our body's essential homeostasis, forcing the system to repeatedly expend energy attempting to regain balance.** Bobba-Alves, Natalia, Robert-Paul Juster, and Martin Picard. 'The Energetic Cost of Allostasis and Allostatic Load'. *Psychoneuroendocrinology* 146 (2022): 105951.

**Self-efficacy refers to the belief in one's ability to execute actions and influence outcomes. Originally proposed by psychologist Albert Bandura, this concept is supported by decades of research across psychology, education, and health.** Bandura, Albert. 'Self-Efficacy: Toward a Unifying Theory of Behavioral Change'. *Psychological Review* 84, no. 2 (1977): 191–215.

**People with high self-efficacy tend to persist through challenges, adapt more effectively to change, and recover more quickly from setbacks.** Benight, Charles, and Roman Cieslak. 'Cognitive Factors and Resilience: How Self-Efficacy Contributes to Coping with Adversities'. In *Resilience and Mental Health: Challenges Across the Lifespan*, edited by Brett Litz, Dennis Charney, Matthew Friedman, and Steven Southwick. Cambridge University Press, 2011.

**It's the internal signal that we are capable of acting in line with our goals, especially under pressure.** Maddux, James 'Self-Efficacy: The Power of Believing You Can'. In *Oxford Handbook of Positive Psychology*, 2nd Ed. Oxford Library of Psychology. Oxford University Press, 2009.

**From an active inference perspective, self-efficacy reflects the system's confidence in its ability to resolve prediction errors and reduce free energy through intentional action. When someone believes they can act effectively, their system is more likely to engage in exploration and adaptive updating, leading to greater coherence across the mind, body, and environment.** Krupnik, Valery. 'I like Therefore I Can, and I Can Therefore I like: The Role of Self-Efficacy and Affect in Active Inference of Allostasis'. *Frontiers in Neural Circuits* 18 (2024): 1283372.

**A common pattern many of us fall into is the overuse of pragmatic willpower, sticking to behaviors we know but that may have diminishing effectiveness, without expanding into epistemic volitions.**

**Navigating Uncertainty**

Maisto, Domenico, Karl Friston, and Giovanni Pezzulo. 'Caching Mechanisms for Habit Formation in Active Inference'. *Neurocomputing* 359 (2019): 298–314; Schwoebel, Sarah, Dimitrije Markovic, and Stefan Kiebel. 'An Active Inference Perspective on Habit Learning'. Paper presented at Conference on Cognitive Computational Neuroscience. *ResearchGate*, 2019.

**While the prefrontal cortex provides direction, the body supports action, and the environment shapes context. By aligning these elements, volition creates a feedback loop where each part reinforces the other.** Frith, Chris. 'The Psychology of Volition'. *Experimental Brain Research* 229, no. 3 (2013): 289–99.

### CHAPTER 15

**Here, embodiment means the body takes an active role in shaping how we think, feel, perceive, and act, rather than merely housing the mind. Our bodily states shape our attention, influence our emotions, and help determine how we interpret the world.** Varela, Francisco, Evan Thompson, and Eleanor Rosch. *The Embodied Mind: Cognitive Science and Human Experience.* MIT Press, 2017.

**By both definition and current research, embodiment isn't something we can achieve through cognition alone. It requires the active participation of the body.** Shapiro, Lawrence A. 'Embodied Cognition.' The Oxford Handbook of Philosophy of Cognitive Science, edited by Eric Margolis  Richard Samuels, and Stephen Stich. Vol. 1. Oxford University Press, 2012.

**Grounding practices can help us reconnect with our cycles, allowing us to process these signals constructively and experience the holistic benefits of being at home within our system. This state of embodiment transforms the body into a sensory organ—an essential source of evidence that complements external inputs like vision.** Price, Cynthia, and Carole Hooven. 'Interoceptive Awareness Skills for Emotion Regulation: Theory and Approach of Mindful Awareness in Body-Oriented Therapy (MABT)'. *Frontiers in Psychology* 9 (2018): 798.

**This process is known as interoceptive inference, using bodily sensations with cognitive signals to maintain balance and a coherent sense of self.** Seth, Anil, and Karl Friston. 'Active Interoceptive Inference and the Emotional Brain'. *Philosophical Transactions of the Royal Society B: Biological Sciences* 371, no. 1708 (2016): 20160007.

**Our body is constantly generating interoceptive signals (heart rate, hunger, muscle tension) which the brain interprets to maintain self-knowing. These signals are not peripheral; they are central to how we experience being "ourselves."** Herbert, Beate, and Olga Pollatos. 'The Body in the Mind: On the Relationship Between Interoception and Embodiment'. *Topics in Cognitive Science* 4, no. 4 (2012): 692–704.

**Instead, emotions arise from the brain's best guesses about the causes of bodily signals, shaped by context and prediction. Increasing the cohesiveness of our body within the active inference cycle sharpens these inputs, making our emotions clearer and more actionable.** Seth, Anil 'Interoceptive Inference, Emotion, and the Embodied Self'. *Trends in Cognitive Sciences* 17, no. 11 (2013): 565–73; Seth, Anil, and Hugo Critchley. 'Extending Predictive Processing to the Body: Emotion as Interoceptive Inference'. *The Behavioral and Brain Sciences* 36, no. 3 (2013): 227–28.

**Embodiment also helps us label our states more accurately.** Parrinello, Noémie, Jessica Napieralski, Alexander Gerlach, and Anna Pohl. 'Embodied Feelings—A Meta-Analysis on the Relation of Emotion Intensity Perception and Interoceptive Accuracy'. *Physiology & Behavior* 254 (2022): 113904.

**Reintegrating these signals into the active inference cycle reveals their connection to states such as stress, turning discomfort into useful sensory information, and empowering us to respond more effectively.** Paulus, Martin, Justin Feinstein, and Sahib Khalsa. 'An Active Inference Approach to Interoceptive Psychopathology'. *Annual Review of Clinical Psychology* 115, no. 1 (2019): 97–122; Quadt, Lisa, Hugo Critchley, and Sarah Garfinkel. 'The Neurobiology of Interoception in Health and Disease'. *Annals of the New York Academy of Sciences* 1428, no. 1 (2018): 112–28.

**Again, we can look at how we tune the precision of these signals by adjusting the confidence and focus of our attention to shape our experiences and outcomes.** Owens, Andrew, Micah Allen, Sasha Ondobaka, and Karl Friston. 'Interoceptive Inference: From Computational Neuroscience to Clinic'. *Neuroscience & Biobehavioral Reviews* 90 (2018): 174–83.

**If our precision is too high, irrelevant signals (like chronic pain without a physical cause) can dominate our experience. If too low, critical cues (like thirst) might go unnoticed.** Ainley, Vivien, Matthew Apps, Aikaterini Fotopoulou, and Manos Tsakiris. '"Bodily Precision": A Predictive Coding Account of Individual Differences in Interoceptive Accuracy'. *Philosophical Transactions of the Royal Society B: Biological Sciences* 371, no. 1708 (2016): 20160003.

**There are numerous activities shown to foster embodiment, including yoga, meditation, somatic practices, breathwork, dance, and immersion in nature.** Corazon, Sus, Theresa Schilhab, and Ulrika Stigsdotter. 'Developing the Therapeutic Potential of Embodied Cognition and Metaphors in Nature-Based Therapy: Lessons from Theory to Practice'. *Journal of Adventure Education and Outdoor Learning* 11, no. 2 (2011): 161–71; Impett, Emily, Jennifer Daubenmier, and Allegra Hirschman. 'Minding the Body: Yoga, Embodiment, and Well-Being'. *Sexuality Research & Social Policy* 3, no. 4 (2006): 39–48.

**Advances in technology, such as immersive virtual reality or sophisticated neural feedback tools, could provide new avenues for deepening embodiment more efficiently through simulated environments or enhanced cognitive training.** Klingenberg, Sara, Robin Bosse, Richard Mayer, and Guido Makransky. 'Does Embodiment in Virtual Reality Boost Learning Transfer? Testing an Immersion-Interactivity Framework'. *Educational Psychology Review* 36, no. 4 (2024): 116.

CHAPTER 16

From an active inference perspective, play serves a valuable purpose: resolving uncertainty about what we can and can't control. Joy can then be understood as our intrinsic signal that emerges when uncertainty is successfully minimized and predictions about the world align with sensory inputs in a meaningful and rewarding way. Andersen, Marc Malmdorf, Julian Kiverstein, Mark Miller, and Andreas Roepstorff. 'Play in Predictive Minds: A Cognitive Theory of Play'. *Psychological Review* 130, no. 2 (2023): 462–79; Kiverstein, Julian, and Mark Miller. 'Playfulness and the Meaningful Life: An Active Inference Perspective'. *Neuroscience of Consciousness* 2023, no. 1 (2023): niad024.

This process of experimentation, observation, and learning builds the foundation of our world model, teaching us cause and effect long before we can articulate it. Babik, Iryna, James Cole Galloway, and Michele Lobo. 'Early Exploration of One's Own Body, Exploration of Objects, and Motor, Language, and Cognitive Development Relate Dynamically across the First Two Years of Life'. *Developmental Psychology* 58, no. 2 (2022): 222–35.

This is especially crucial in early development, where children begin to grasp the concept of "self" and "other." Before we can understand ourselves, we must first recognize that others exist independently of us. Simantov, Tslil, Michael Lombardo, Simon Baron-Cohen, and Florina Uzefovsky. 'Self-Other Distinction'. *The Neural Basis of Mentalizing* (2021): 85-106; Steinbeis, Nikolaus. 'The Role of Self–Other Distinction in Understanding Others' Mental and Emotional States: Neurocognitive Mechanisms in Children and Adults'. *Philosophical Transactions of the Royal Society B: Biological Sciences* 371, no. 1686 (2016): 20150074.

It offers space to explore the boundaries between self and other, sharpen a sense of agency, and engage with the subtleties of social interaction. Weisberg, Deena Skolnick. 'Pretend Play'. *WIREs Cognitive Science* 6, no. 3 (2015): 249–61.

Moments of shared play help build trust and mutual understanding, reinforcing connection and collaboration. Jarvis, Pam, Stephen Newman, and Louise Swiniarski. 'On "Becoming Social": The Importance of Collaborative Free Play in Childhood'. *International Journal of Play* 3, no. 1 (2014): 53–68.

At the neurological level, play activates the brain's reward system, releasing chemicals that enhance learning, bonding, and motivation. Trezza, Viviana, Petra Baarendse, and Louk Vanderschuren. 'The Pleasures of Play: Pharmacological Insights into Social Reward Mechanisms'. *Trends in Pharmacological Sciences* 31, no. 10 (2010): 463–69.

Cultural expectations often prioritize productivity over leisure, psychological barriers such as fear of judgment or failure can create resistance, and practical challenges like demanding schedules and responsibilities can feel like they leave little time for play. Van Vleet, Meredith, and Brooke Feeney. 'Play Behavior and Playfulness in Adulthood'. *Social and Personality Psychology Compass* 9, no. 11 (2015): 630–43.

That's why play is now being used in therapeutic contexts, from video games that offer safe spaces for feedback and experimentation, to role-playing games like Dungeons & Dragons that help people explore emotional dynamics through story and imagination. Griffiths, Mark 'The Therapeutic and Health Benefits of Playing Video Games'. In *The Oxford Handbook of Cyberpsychology*, edited by Alison Attrill-Smith, Chris Fullwood, Melanie Keep, and Daria Kuss. Oxford University Press, 2019; Henrich, Sören, and Rachel Worthington. 'Let Your Clients Fight Dragons: A Rapid Evidence Assessment Regarding the Therapeutic Utility of "Dungeons & Dragons"'. *Journal of Creativity in Mental Health* 18, no. 3 (2023): 383–401.

In other words, play updates our active inference cycles and reduces free energy. Whether we are developing new skills, engaging in creative exploration, or strengthening relationships, play provides a powerful avenue for deeper self-discovery and a greater sense of agency in the world. Kiverstein, Julian, and Mark Miller. 'Playfulness and the Meaningful Life: An Active Inference Perspective'. *Neuroscience of Consciousness* 2023, no. 1 (2023): niad024.

## CHAPTER 17

Scientifically, this is a personalized example of experiencing the process of mentalizing interoception—becoming skilled at interpreting the signals the body sends and integrating them into your sense of self. Khalsa, Sahib, Ralph Adolphs, Oliver Cameron, et al. 'Interoception and Mental Health: A Roadmap'. *Biological Psychiatry. Cognitive Neuroscience and Neuroimaging* 3, no. 6 (2018): 501–13; Musculus, Lisa, Markus R. Tünte, Markus Raab, and Ezgi Kayhan. 'An Embodied Cognition Perspective on the Role of Interoception in the Development of the Minimal Self'. *Frontiers in Psychology* 12 (2021): 716950.

Play fixtures such as meditation practices that went far beyond typical mindfulness apps, somatic and body-based work well outside mainstream yoga studios, group dynamics that extend past standard improv exercises, and cultural rituals that carry embodied wisdom from generations. Ma-Kellams, Christine. 'Cross-Cultural Differences in Somatic Awareness and Interoceptive Accuracy: A Review of the Literature and Directions for Future Research'. *Frontiers in Psychology* 5 (2014): 1379.

## CHAPTER 18

However, from the perspective of active inference and self-evidencing, [Definition process by which our brains continuously seek to maximize the evidence for their internal models of the world. Essentially, the brain is constantly testing its predictions against sensory input to ensure that its understanding of the environment is as accurate and useful as possible] rest plays a much more complex and essential role in how we navigate the world and refine our internal models. Hohwy, Jakob. 'Conscious Self-Evidencing'. *Review of Philosophy and Psychology* 13, no. 4 (2022): 809–28.

It helps strike a balance between two key components of self-evidencing: accuracy (how well our internal models fit our sensory perceptions) and complexity, which refers to the degree of detail in our models. Smith, Ryan, Karl Friston, and Christopher Whyte. 'A Step-by-Step Tutorial on Active Inference and Its Application to Empirical Data'. *Journal of Mathematical Psychology* 107 (2022): 102632.

If our models become too complex, we risk overfitting and adding unnecessary details that hinder adaptability... Conversely, if our models are too simple, we risk underfitting and failing to capture important nuances of our experiences. Venter, Elmarie. 'Toward an Embodied, Embedded Predictive Processing Account'. *Frontiers in Psychology* 12 (2021): 543076.

Rest is a key method to help us arrive at this balance as we navigate life. It's like a form of cognitive housekeeping, allowing us to refine our models by simplifying them without losing essential information. Hoel, Erik. 'The Overfitted Brain: Dreams Evolved to Assist Generalization'. *Patterns* 2, no. 5 (2021): 100244.

As Einstein famously put it, we aim to "keep everything as simple as possible, but no simpler." Einstein, Albert. 'On the method of theoretical physics.' *Philosophy of science* 1, no. 2 (1934): 163-169; Robinson, Andrew. 'Did Einstein Really Say That?' *Nature* 557, no. 7703 (2018): 30–1.

During sleep, our brain engages in a crucial process known as synaptic homeostasis, a concept proposed by Giulio Tononi. This process involves pruning unnecessary neural connections while also strengthening those most relevant to learning and memory, leaving behind a refined, simplified structure that is better suited for the challenges of the next day. Tononi, Giulio, and Chiara Cirelli. 'Sleep and the Price of Plasticity: From Synaptic and Cellular Homeostasis to Memory Consolidation and Integration'. *Neuron* 81, no. 1 (2014): 12–34; Tononi, Giulio, and Chiara Cirelli. 'Sleep Function and Synaptic Homeostasis'. *Sleep Medicine Reviews* 10, no. 1 (2006): 49–62.

When we disengage from sensory input, such as closing our eyes for a moment, staring off into the distance, or engaging in quiet reflection, we create space to reorganize our thoughts and avoid cognitive overload. This momentary disengagement prevents the overfitting of our internal models by offering a chance to recalibrate and prioritize what is truly relevant. Jebelli, Joseph. The Brain at Rest. 2025; Weng, Linman, Jing Yu, Zhangwei Lv, Shiyan Yang, Simon Theodor Jülich, and Xu Lei. 'Effects of Wakeful Rest on Memory Consolidation: A Systematic Review and Meta-Analysis'. *Psychonomic Bulletin & Review* 32, no. 5 (2025): 1937–68.

Yet it exists across species, suggesting its fundamental importance. Siegel, Jerome 'Sleep Function: An Evolutionary Perspective'. *The Lancet Neurology* 21, no. 10 (2022): 937–46.

Rest stands as a fundamental aspect of life, shaping the development of adaptive, well-tuned models that guide decisions and sustain resilience over time. Suchecki, Deborah, Paula Tiba, and Ricardo Machado. 'REM Sleep Rebound as an Adaptive Response to Stressful Situations'. *Frontiers in Neurology* 3 (2012): 41; Zimmerman, Molly, Giada Benasi, Christiane Hale, et al. 'The Effects of Insufficient Sleep and Adequate Sleep on Cognitive Function in Healthy Adults'. *Sleep Health* 10, no. 2 (2024): 229–36.

**References**

Incorporating intentional rest, whether through sleep, meditation, or moments of quiet reflection, can help us:

- Enhance learning and memory integration
- Improve decision-making and problem-solving
- Reduce mental fatigue and overwhelm
- Strengthen emotional resilience and well-being

McElroy, Todd, and David Dickinson. 'Thinking about Complex Decisions: How Sleep and Time-of-Day Influence Complex Choices'. *Consciousness and Cognition* 76 (2019): 102824; Salfi, Federico, Marco Lauriola, Daniela Tempesta, et al. 'Effects of Total and Partial Sleep Deprivation on Reflection Impulsivity and Risk-Taking in Deliberative Decision-Making'. *Nature and Science of Sleep* 12 (2020): 309–24; Sanders, Kristin, Samuel Osburn, Ken A. Paller, and Mark Beeman. 'Targeted Memory Reactivation During Sleep Improves Next-Day Problem Solving'. *Psychological Science* 30, no. 11 (2019): 1616–24; Tomaso, Cara, Anna Johnson, and Timothy Nelson. 'The Effect of Sleep Deprivation and Restriction on Mood, Emotion, and Emotion Regulation: Three Meta-Analyses in One'. *Sleep* 44, no. 6 (2021): zsaa289; Weinberg, Melissa, Jacqueline Noble, and Thomas Hammond. 'Sleep Well Feel Well: An Investigation into the Protective Value of Sleep Quality on Subjective Well-being'. *Australian Journal of Psychology* 68, no. 2 (2016): 91–97.

**CHAPTER 19**

**It feels as though my limbic system (part of my larger threat and arousal systems) is extending its tentacles across my consciousness, heightening my senses and pulling my attention into a more aware state, readying itself to respond to what's happening around me.** Bertram, Teresa, Daniel Hoffmann Ayala, Maria Huber, et al. 'Human Threat Circuits: Threats of Pain, Aggressive Conspecific, and Predator Elicit Distinct BOLD Activations in the Amygdala and Hypothalamus'. *Frontiers in Psychiatry* 13 (2023): 1063238.

**Pain, especially chronic pain, is a complex phenomenon influenced by environmental stressors, mental constructs, and physical sensations.** Mills, Sarah, Karen Nicolson, and Blair Smith. 'Chronic Pain: A Review of Its Epidemiology and Associated Factors in Population-Based Studies'. *British Journal of Anaesthesia* 123, no. 2 (2019): e273-e283.

**This phenomenon, in part, can be explained by the brain's plasticity—the way it adapts and rewires itself based on repeated patterns.** Mansour, A.R., M.A. Farmer, M.N. Baliki, and A. Vania Apkarian. 'Chronic Pain: The Role of Learning and Brain Plasticity'. *Restorative Neurology and Neuroscience* 32, no. 1 (2014): 129–39.

**Pain, especially chronic pain, involves more than just physical damage; it becomes embedded in the brain's predictive coding system. The brain learns to associate certain sensations with danger or injury, even when the original cause has healed, perpetuating the experience of pain.** Castejón, Jorge, Feifan Chen, Anusha Yasoda-Mohan, Colum Ó Sé, and Sven Vanneste. 'Chronic Pain – A Maladaptive Compensation to Unbalanced

Hierarchical Predictive Processing'. *NeuroImage* 297 (2024): 120711; Hechler, Tanja, Dominik Endres, and Anna Thorwart. 'Why Harmless Sensations Might Hurt in Individuals with Chronic Pain: About Heightened Prediction and Perception of Pain in the Mind'. *Frontiers in Psychology* 7 (2016): 1638; Kiverstein, Julian, Michael D. Kirchhoff, and Mick Thacker. 'An Embodied Predictive Processing Theory of Pain Experience'. *Review of Philosophy and Psychology* 13, no. 4 (2022): 973–98.

**The brain's predictive coding system, which continuously updates models of the body and environment to reduce surprise, can at times reinforce unhelpful patterns.** Qela, Brendon, Stefano Damiani, Samanta De Santis, et al. 'Predictive Coding in Neuropsychiatric Disorders: A Systematic Transdiagnostic Review'. *Neuroscience & Biobehavioral Reviews* 169 (2025): 106020.

**Researchers like Andy Clark highlight how cognitive constraints and environmental factors shape these processes, emphasizing that individuals are doing their best within the limits of their resources and context.** Butz, Martin, Maximilian Mittenbühler, Sarah Schwöbel, et al. 'Contextualizing Predictive Minds'. *Neuroscience & Biobehavioral Reviews* 168 (2025): 105948; Clark, Andy. *Surfing Uncertainty: Prediction, Action, and the Embodied Mind.* Oxford University Press, 2015; Clark, Andy. 'Whatever next? Predictive Brains, Situated Agents, and the Future of Cognitive Science'. *Behavioral and Brain Sciences* 36, no. 3 (2013): 181–204.

**Crucially, the brain's adaptability (and its plasticity) also offers hope, providing opportunities to recalibrate and rewire these cycles toward healing and growth.** Gazerani, Parisa. 'The Neuroplastic Brain: Current Breakthroughs and Emerging Frontiers'. *Brain Research* 1858 (2025): 149643; Kays, Jill, Robin Hurley, and Katherine Taber. 'The Dynamic Brain: Neuroplasticity and Mental Health'. *The Journal of Neuropsychiatry and Clinical Neurosciences* 24, no. 2 (2012): 118–24.

**We get to choose which sensations to engage with, what to mentalize, how much confidence to assign them, and how to practice and learn. This agency allows us to craft experiences that contain more ease and harmony, even when discomfort arises.** Friston, Karl, Philipp Schwartenbeck, Thomas Fitzgerald, Michael Moutoussis, Tim Behrens, and Raymond Dolan. 'The Anatomy of Choice: Active Inference and Agency'. *Frontiers in Human Neuroscience* 7 (2013): 598.

**Pain Reprocessing Therapy (PRT) is an emerging approach that teaches the brain to reinterpret pain signals. A 2021 randomized controlled trial in JAMA Psychiatry showed significant pain relief in chronic back pain patients using this method.** Ashar, Yoni, Alan Gordon, Howard Schubiner, et al. 'Effect of Pain Reprocessing Therapy vs Placebo and Usual Care for Patients With Chronic Back Pain: A Randomized Clinical Trial'. *JAMA Psychiatry* 79, no. 1 (2022): 13–23; Unwinding Anxiety®. 'Unwinding Anxiety® - Anxiety Management & Relief Program'. Accessed 19 September 2025. https://unwindinganxiety.com/.

The Unwinding Anxiety App by Dr. Judson Brewer uses mindfulness and reward-based learning to help break the habit loops that drive anxiety and physiological overactivation. *Pain Reprocessing Therapy (PRT) | Break Free from Chronic Pain.* Accessed 19 September 2025. https://www.painreprocessingtherapy.com/.

**CHAPTER 20**

**The inner critic metaphorically narrows our prediction landscape, reinforcing expectations of failure or inadequacy. Self-compassion, on the other hand, softens rigid predictions and allows for more adaptive learning.** Diedrich, Alice, Michaela Grant, Stefan Hofmann, Wolfgang Hiller, and Matthias Berking. 'Self-Compassion as an Emotion Regulation Strategy in Major Depressive Disorder'. *Behaviour Research and Therapy* 58 (2014): 43–51; Foroughi, Aliakbar, Kheirollah Sadeghi, AliAkbar Parvizifard, et al. 'The Effectiveness of Mindfulness-Based Cognitive Therapy for Reducing Rumination and Improving Mindfulness and Self-Compassion in Patients with Treatment-Resistant Depression'. *Trends in Psychiatry and Psychotherapy* 42 (2020): 138–46; Pedro, Liliana, Mariana Branquinho, Maria Cristina Canavarro, and Ana Fonseca. 'Self-Criticism, Negative Automatic Thoughts and Postpartum Depressive Symptoms: The Buffering Effect of Self-Compassion'. *Journal of Reproductive and Infant Psychology* 37, no. 5 (2019): 539–53.

**When we assign rigid labels to experiences, such as "bad," "wrong," "a failure," we reinforce rigid priors, which makes it harder to use prediction errors adaptively for recalibration.** Lane, Richard, and Ryan Smith. 'Levels of Emotional Awareness: Theory and Measurement of a Socio-Emotional Skill'. *Journal of Intelligence* 9, no. 3 (2021): 42; Qela, Brendon, Stefano Damiani, Samanta De Santis, et al. 'Predictive Coding in Neuropsychiatric Disorders: A Systematic Transdiagnostic Review'. *Neuroscience & Biobehavioral Reviews* 169 (2025): 106020; Smith, Ryan, Paul Badcock, and Karl Friston. 'Recent Advances in the Application of Predictive Coding and Active Inference Models within Clinical Neuroscience'. *Psychiatry and Clinical Neurosciences* 75, no. 1 (2021): 3–13.

**When we grasp for excessive control, we resist the natural unfolding of experience, failing to reduce free energy effectively by fighting against uncertainty.** Fradkin, Isaac, Rick Adams, Thomas Parr, Jonathan Roiser, and Jonathan Huppert. 'Searching for an Anchor in an Unpredictable World: A Computational Model of Obsessive Compulsive Disorder.' *Psychological Review* 127, no. 5 (2020): 672–99.

**Reactivity often arises from deeply ingrained priors as automatic responses that may no longer serve us.** Harris, Hebert 'Active Inference and Psychodynamics: A Novel Integration with Applications to Depression and Stress Disorders'. *Frontiers in Psychiatry* 16 (2025): 1630858; Friston, Karl, Thomas FitzGerald, Francesco Rigoli, Philipp Schwartenbeck, and Giovanni Pezzulo. "Active inference and learning." *Neuroscience & Biobehavioral Reviews* 68 (2016): 862-879.

**Inaction is a form of action; rest and recovery help maintain physiological and mental balance by activating the parasympathetic nervous system and supporting homeostasis.** John Hall. Guyton and Hall

Textbook of Medical Physiology. 13th ed. Elsevier Health Sciences, 2015.

**Stress and fear tend to shrink our perceptual field, limiting possible actions by heightening sympathetic arousal and narrowing attention to perceived threats.** Arnaldo, Irene, Andrew Corcoran, Karl Friston, and Maxwell Ramstead. 'Stress and Its Sequelae: An Active Inference Account of the Etiological Pathway from Allostatic Overload to Depression'. *Neuroscience & Biobehavioral Reviews* 135 (2022): 104590; Grogans, Shannon, Eliza Bliss-Moreau, Kristin Buss, et al. 'The Nature and Neurobiology of Fear and Anxiety: State of the Science and Opportunities for Accelerating Discovery'. *Neuroscience & Biobehavioral Reviews* 151 (2023): 105237.

**Expansion through openness, curiosity, and presence activates broader networks in the brain, allowing for more flexible predictions and adaptive action in response to new possibilities.** Brewer, Judson, and Fabio Giommi. 'Psychotherapy as Investigation: Cultivating Curiosity and Insight in the Therapeutic Process'. *Frontiers in Psychology* 16 (2025): 1603719.

**Avoiding discomfort can reinforce maladaptive predictions by signaling to the brain that the sensation is dangerous, keeping us stuck in avoidance loops.** Brown, Vanessa, Rebecca Price, and Alexandre Dombrovski. 'Anxiety as a Disorder of Uncertainty: Implications for Understanding Maladaptive Anxiety, Anxious Avoidance, and Exposure Therapy'. *Cognitive, Affective & Behavioral Neuroscience* 23, no. 3 (2023): 844–68; Flores, Amanda, Francisco López, Bram Vervliet, and Pedro Cobos. 'Intolerance of Uncertainty as a Vulnerability Factor for Excessive and Inflexible Avoidance Behavior'. *Behaviour Research and Therapy* 104 (2018): 34–43; Kube, Tobias, Max Berg, Birgit Kleim, and Philipp Herzog. 'Rethinking Post-Traumatic Stress Disorder – A Predictive Processing Perspective'. *Neuroscience & Biobehavioral Reviews* 113 (2020): 448–60.

**Engaging with sensation (even discomfort) in a measured way allows for recalibration by updating the brain's predictions based on new, safe experiences.** Smith, Ryan, Michael Moutoussis, and Edda Bilek. 'Simulating the Computational Mechanisms of Cognitive and Behavioral Psychotherapeutic Interventions: Insights from Active Inference'. *Scientific Reports* 11, no. 1 (2021): 10128.

**Fixating on rigid interpretations of experience limits adaptability by sustaining prediction errors and leaving them unresolved.** Smith, Ryan, Lav Varshney, Susumu Nagayama, Masahiro Kazama, Takuya Kitagawa, and Yoshiki Ishikawa. 'A Computational Neuroscience Perspective on Subjective Wellbeing within the Active Inference Framework'. *International Journal of Wellbeing* 12, no. 4 (2022); Sopp, Roxanne, Shilat Haim-Nachum, Moritz Braun, Johanna Lass-Hennemann, Sarah Schäfer, and Tanja Michael. 'How We See the World: Inflexible Interpretation Updating as a Predictor and Moderator of PTSD Symptoms in High-Risk Occupations'. *Psychological Trauma: Theory, Research, Practice and Policy*, ahead of print, 9 January 2025.

**Seeking an immediate resolution to discomfort can prevent deeper learning by reinforcing the brain's urgency to eliminate uncertainty rather than understand it. Letting experiences integrate over time**

allows for a more sustainable recalibration, as new neural pathways form through gradual meaning-making and prediction updating. Van De Cruys, Sander. 'Affective Value in the Predictive Mind'. *Philosophy and Predictive Processing*, Theoretical Philosophy/MIND Group – JGU Mainz, 2017; Voss, Patrice, Maryse Thomas, Miguel Cisneros-Franco, and Étienne de Villers-Sidani. 'Dynamic Brains and the Changing Rules of Neuroplasticity: Implications for Learning and Recovery'. *Frontiers in Psychology* 8 (2017): 274878.

Rumination, cycling through the same thoughts again and again, keeps us stuck by reinforcing existing predictions rather than allowing for flexible adaptation. Davey, Christopher 'The Body Intervenes: How Active Inference Explains Depression's Clinical Presentation'. *Neuroscience & Biobehavioral Reviews* 175 (2025): 106229.

## CHAPTER 21

Across cultures, religions, and philosophies, love has been regarded as a central force in morality, a fundamental source of meaning, and a key contributor to fulfillment. Karandashev, Victor. *Cross-Cultural Perspectives on the Experience and Expression of Love*. Springer International Publishing, 2019; Pismenny, Arina, and Berit Brogaard, eds. *The Moral Psychology of Love*. Rowman & Littlefield Publishers, 2022.

Christianity speaks of agape, an unconditional love that transcends the self, while Judaism emphasizes ahavah, a love rooted in covenant, justice, and ethical responsibility, unconditional love that transcends the self. Buddhism teaches metta, or loving-kindness, as a practice for dissolving barriers between oneself and others. Hinduism embraces bhakti, devotional love, as a path to the divine. In Islam, love is at the heart of one's relationship with God and the community. Greenberg, Yudit Kornberg. *Encyclopedia of Love in World Religions*: ABC-CLIO, 2008; Nakissa, Aria. 'Comparing Moralities in the Abrahamic and Indic Religions Using Cognitive Science: Kindness, Peace, and Love versus Justice, Violence, and Hate'. *Religions* 14, no. 2 (2023): 203.

Greek, for instance, distinguishes between eros (romantic love), philia (friendship), and agape (universal, unconditional love). Rinne, Pärttyli, Mikke Tavast, Enrico Glerean, and Mikko Sams. 'Body Maps of Loves'. *Philosophical Psychology* 38, no. 4 (2025): 1453–75.

If we think about love from an active inference perspective, we can see it's fundamental to our manifestation of our cycles because it expands our sense of self not only physically but cognitively as well. This happens as our brain integrates the other person into both our bodily regulation and our internal model of the world. Friston, Karl, and Christopher Frith. 'Active Inference, Communication and Hermeneutics'. *Cortex; a Journal Devoted to the Study of the Nervous System and Behavior* 68 (2015): 129–43.

Whether romantic, familial, platonic, or altruistic, cognitively it acts as a sense-making process [definition: the process by which people give meaning to their collective experiences], shaping how we engage with the world and with one another. It fosters prosocial behaviors, including emotional

**resilience, strengthens social bonds, and enhances our capacity for joy.** Brown, Casey, and Barbara Fredrickson. 'Characteristics and Consequences of Co-Experienced Positive Affect: Understanding the Origins of Social Skills, Social Bonds, and Caring, Healthy Communities'. *Current Opinion in Behavioral Sciences* 39 (2021): 58–63; Cheadle, Jacob, Davidson-Turner, and Bridget Goosby. 'Active Inference and Social Actors: Towards a Neuro-Bio-Social Theory of Brains and Bodies in Their Worlds'. *KZfSS Kölner Zeitschrift Für Soziologie Und Sozialpsychologie* 76, no. 3 (2024): 317–50; Moutoussis, Michael, Nelson Jesús Trujillo-Barreto, Wael El-Deredy, Raymond Dolan, and Karl Friston. 'A Formal Model of Interpersonal Inference'. *Frontiers in Human Neuroscience* 8 (2014): 160.

**Romantic love, when viewed through the lens of active inference, is not merely an experience but a deep, embodied inference. It includes subtle cues, such as tones, gestures, and postures that our body often picks up before our conscious mind does.** Ortigue, Stephanie, Nisa Patel, Francesco Bianchi-Demicheli, and Scott Grafton. 'Implicit Priming of Embodied Cognition on Human Motor Intention Understanding in Dyads in Love'. *Journal of Social and Personal Relationships* 27, no. 7 (2010): 1001–15.

**Love, as a generative or mental model process, integrates bodily sensations and social interactions, forming our ability to connect and build relationships. When we engage with others, we update these generative models.** Veissière, Samuel, Axel Constant, Maxwell Ramstead, Karl Friston, and Laurence Kirmayer. 'Thinking through Other Minds: A Variational Approach to Cognition and Culture'. *Behavioral and Brain Sciences* 43 (2020): e90.

**Across these various bonds, sensory cues function as essential signals that continuously update our understanding of others and enhance our emotional attunement.** Prior, Nora, Ehren Bentz, and Alexander Ophir. 'Reciprocal Processes of Sensory Perception and Social Bonding: An Integrated Social-Sensory Framework of Social Behavior'. *Genes, Brain and Behavior* 21, no. 3 (2022): e12781; Decety, Jean, and Philip Jackson. 'The functional architecture of human empathy. *Behavioral and cognitive neuroscience reviews* 3, no. 2 (2004): 71-100.

**Building on this, love moves beyond individual awareness and unfolds as a continuous cycle of mutual inference. When two people care about each other, they actively adjust their internal models in response to one another's actions, emotions, and expressions.** Lehmann, Konrad, Dimitris Bolis, Karl Friston, Leonhard Schilbach, Maxwell Ramstead, and Philipp Kanske. 'An Active-Inference Approach to Second-Person Neuroscience'. *Perspectives on Psychological Science* 19, no. 6 (2024): 931–51.

**Self-love, from an active inference perspective, is the ongoing process of updating our beliefs about ourselves with kindness and curiosity.** Neff, Kristin. 'Self-compassion: An Alternative Conceptualization of a Healthy Attitude Toward Oneself.' *Self and identity* 2, no. 2 (2003): 85-101.

**It increases our ability to regulate stress, make healthier decisions, and remain resilient in relationships. Without self-love, we are more prone to rigid beliefs, reactive behaviors, and chronic states of**

**References**

**physiological or emotional dysregulation.** Ardi, Ziv, Yulia Golland, Roni Shafir, Gal Sheppes, and Nava Levit-Binnun. 'The Effects of Mindfulness-Based Stress Reduction on the Association Between Autonomic Interoceptive Signals and Emotion Regulation Selection'. *Psychosomatic Medicine* 83, no. 8 (2021): 852–62; Fields, Eric, and Gina Kuperberg. 'Loving Yourself More than Your Neighbor: ERPs Reveal Online Effects of a Self-Positivity Bias'. *Social Cognitive and Affective Neuroscience* 10, no. 9 (2015): 1202–9; Klussman, Kristine, Nicola Curtin, Julia Langer, and Austin Lee Nichols. 'The Importance of Awareness, Acceptance, and Alignment With the Self: A Framework for Understanding Self-Connection'. *Europe's Journal of Psychology* 18, no. 1 (2022): 120–31.

**Individuals in close relationships increasingly integrate their loved ones into their predictive models, leading to a greater sense of shared identity.** Aron, Arthur, Gary Lewandowski, Brittany Branand, Debra Mashek, and Elaine Aron. 'Self-Expansion Motivation and Inclusion of Others in Self: An Updated Review'. *Journal of Social and Personal Relationships* 39, no. 12 (2022): 3821–52; Cruwys, Tegan, Erica South, William Kim Halford, Judith Murray, and Martin Fladerer. 'Measuring "We-Ness" in Couple Relationships: A Social Identity Approach'. *Family Process* 62, no. 2 (2023): 795–817.

**This ability to see oneself as part of a larger system aligns with broader ideas of social cohesion and altruism, but it is also inherently self-serving.** Ansbacher, Heinz, and Rowena Ansbacher, eds. *The Individual Psychology of Alfred Adler.* HarperCollins, 1964.

**Love enhances emotional resilience, reduces stress, and improves overall well-being.** Hojjat, Mahzad, and Duncan Cramer, eds. Positive Psychology of Love. Positive Psychology of Love. Oxford University Press, 2013; Kahana, Eva, Tirth Bhatta, Boaz Kahana, and Nirmala Lekhak. 'Loving Others: The Impact of Compassionate Love on Later-Life Psychological Well-Being'. *The Journals of Gerontology: Series B* 76, no. 2 (2021): 391–402.

**Even for those who are not motivated by altruism, love offers practical benefits: increased cooperation, greater efficiency in collaboration, and an enriched sense of purpose.** Aron, Arthur, and Elaine Aron. 'The Meaning of Love'. *In The Human Quest for Meaning*, 2nd ed. Routledge, 2012; Kleef, Gerben van, and Gert-Jan Lelieveld. 'Moving the Self and Others to Do Good: The Emotional Underpinnings of Prosocial Behavior'. *Current Opinion in Psychology* 44 (2022): 80–88.

**The process of 'not like me' becomes even more concerning when it moves from simple distance to dehumanization. [Definition: the process of depriving a person or group of positive human qualities]. When dehumanization happens we stop engaging, stop updating, and over time, stop perceiving.** Kteily, Nour, and Alexander Landry. 'Dehumanization: Trends, Insights, and Challenges'. *Trends in Cognitive Sciences* 26, no. 3 (2022): 222–40.

**But active inference tells us something deeper: when we stop gathering new evidence, we cut ourselves off from the very signals that could revise our beliefs. Ultimately, we become trapped when error goes**

**Navigating Uncertainty**

unquestioned and curiosity to re-examine our beliefs is absent. Friston, Karl, Francesco Rigoli, Dimitri Ognibene, Christoph Mathys, Thomas Fitzgerald, and Giovanni Pezzulo. 'Active Inference and Epistemic Value'. *Cognitive Neuroscience* 6, no. 4 (2015): 187–214.

As we've seen, rigid prior beliefs about someone can prevent us from updating in light of new evidence about who they are. The same applies when past experiences shape our expectations with others in the present, causing us to overweight familiar outcomes and overlook evidence that doesn't fit. Albarracin, Mahault, Daphne Demekas, Maxwell Ramstead, and Conor Heins. 'Epistemic Communities under Active Inference'. *Entropy* 24, no. 4 (2022): 476; Kube, Tobias, and Liron Rozenkrantz. 'When Beliefs Face Reality: An Integrative Review of Belief Updating in Mental Health and Illness'. *Perspectives on Psychological Science: A Journal of the Association for Psychological Science* 16, no. 2 (2021): 247–74.

We can explore practices like forgiveness [In this case, we could see forgiveness as the process of loosening the grip of old predictions shaped by past pain, not a forced bypass], reprocessing past pain, or revisiting old assumptions to further reduce the weight of outdated predictions that may otherwise distort our social perception. Akhtar, Sadaf, and Jane Barlow. 'Forgiveness Therapy for the Promotion of Mental Well-Being: A Systematic Review and Meta-Analysis'. *Trauma, Violence, & Abuse* 19, no. 1 (2018): 107–22.

Practices like meditation are another tool we can use to foster deeper love in ourselves and with others. It enhances our ability to recognize and regulate our internal states, improving our capacity to connect. Petrovic, Julia, Jessica Mettler, Sohyun Cho, and Nancy Heath. 'The Effects of Loving-Kindness Interventions on Positive and Negative Mental Health Outcomes: A Systematic Review and Meta-Analysis'. *Clinical Psychology Review* 110 (2024): 102433.

Through greater attentional control and awareness of interoceptive signals, individuals can become more attuned to their bodily cues, such as heart rate changes, breath patterns, or muscle tension. These often serve as unconscious indicators of emotional states. Ardi, Ziv, Yulia Golland, Roni Shafir, Gal Sheppes, and Nava Levit-Binnun. 'The Effects of Mindfulness-Based Stress Reduction on the Association Between Autonomic Interoceptive Signals and Emotion Regulation Selection'. *Psychosomatic Medicine* 83, no. 8 (2021): 852–62; Lazzarelli, Alessandro, Francesca Scafuto, Cristiano Crescentini, et al. 'Interoceptive Ability and Emotion Regulation in Mind–Body Interventions: An Integrative Review'. *Behavioral Sciences* 14, no. 11 (2024): 1107.


**CHAPTER 22**

In active inference, safety lowers the cost of uncertainty, making greater exploration possible. Villiger, Daniel. 'An Integrative Model of Psychotherapeutic Interventions Based on a Predictive Processing Framework'. *Journal of Contemporary Psychotherapy* 55, no. 1 (2025): 39–49.

**References**

**Self-love eases this rigidity, allowing emotional truths and evolving preferences to be integrated more flexibly.** Gutiérrez-Hernández, María Elena, Luisa Fernanda Fanjul Rodríguez, Alicia Díaz Megolla, Cristián Oyanadel, and Wenceslao Peñate Castro. 'The Effect of Daily Meditative Practices Based on Mindfulness and Self-Compassion on Emotional Distress under Stressful Conditions: A Randomized Controlled Trial'. *European Journal of Investigation in Health, Psychology and Education* 13, no. 4 (2023): 762–75.

**Active inference also depends on reading our internal signals clearly.** Farb, Norman, Jennifer Daubenmier, Cynthia Price, et al. 'Interoception, Contemplative Practice, and Health'. *Frontiers in Psychology* 6 (2015): 763.

**Self-love increases our opportunity to feel what's truly present, such as grief, vulnerability, or longing, without suppressing or distorting it.** Abramson, Kate, and Adam Leite. 'Self-Love and Self-Acceptance'. In *The Philosophy and Psychology of Ambivalence*. Routledge, 2020.

**Freud's model of the mind introduced the primary process, associated with instinctual, emotion-driven thought (the id), and the secondary process, which is logical, structured, and tied to our sense of self (the ego).** Freud, Sigmund. *The Standard Edition of the Complete Psychological Works of Sigmund Freud*. The Standard Edition of the Complete Psychological Works of Sigmund Freud, edited by James Strachey. Macmillan, 1964.

**One way to read Freud's distinction, in light of modern neuroscience, is as a hierarchical inference machine between these two processes. Meaning the mind continuously generates predictions, evaluates sensory input, and updates its internal model to minimize the difference between the two.** Holmes, Jeremy. 'Friston, Free Energy, and Psychoanalytic Psychotherapy'. Entropy 26, no. 4 (2024): 343; Holmes, Jeremy. 'Friston's Free Energy Principle: New Life for Psychoanalysis?' *BJPsych Bulletin* 46, no. 3 (2022): 164–68.

**The primary process aligns with id-driven thinking, characterized by free-associative, imaginative, and emotion-driven cognition. The secondary process, or ego-driven thinking, introduces structure, suppressing chaotic impulses and refining thought into a coherent narrative aligned with external reality.** Boag, Simon. 'Ego, Drives, and the Dynamics of Internal Objects'. *Frontiers in Psychology* 5 (2014): 666.

**These two processes are constantly active, running in parallel and exchanging information beneath our conscious awareness.** Solms, Mark, and Jaak Panksepp. 'The "Id" Knows More than the "Ego" Admits: Neuropsychoanalytic and Primal Consciousness Perspectives on the Interface between Affective and Cognitive Neuroscience'. *Brain Sciences* 2, no. 2 (2012): 147–75.

**Our brain constantly works to minimize the uncertainty (or free energy) between what we expect and what we experience between these two levels.** Friston, Karl. 'The Free-Energy Principle: A Rough Guide to the Brain?' *Trends in Cognitive Sciences* 13, no. 7 (2009): 293–301.

A key player in this balancing act is the Default Mode Network (DMN), a large-scale brain network that engages in self-referential thought, memory, and identity formation. Menon, Vinod. '20 Years of the Default Mode Network: A Review and Synthesis'. *Neuron* 111, no. 16 (2023): 2469–87.

The DMN plays a central role in integrating the primary and secondary processes, allowing us to maintain a coherent self while navigating the world. Cieri, Filippo, and Roberto Esposito. 'Psychoanalysis and Neuroscience: The Bridge Between Mind and Brain'. *Frontiers in Psychology* 10 (2019): 1790.

However, when this system is disrupted, it can lead to dream-like thinking in waking life, which manifests as intrusive thoughts, flashbacks, or dissociation. Viard, Armelle, Justine Mutlu, Sandra Chanraud, et al. 'Altered Default Mode Network Connectivity in Adolescents with Post-Traumatic Stress Disorder'. *NeuroImage: Clinical* 22 (2019): 101731.

If the brain cannot effectively suppress chaotic signals from lower levels, it leads to heightened free energy and, consequently, greater stress and dissonance. Carhart-Harris, Robin, and Karl Friston. 'The Default-Mode, Ego-Functions and Free-Energy: A Neurobiological Account of Freudian Ideas'. *Brain* 133, no. 4 (2010): 1265–83.

For example, suppressing unresolved trauma requires continuous effort to keep distressing memories from surfacing, leading to chronic tension and heightened emotional reactivity. Maté, Dr Gabor. *When the Body Says No: The Cost of Hidden Stress*. Vermilion, 2019.

Substance abuse artificially alters perception, forcing the brain to work harder to reconcile distorted signals with reality, which can result in cognitive fatigue and dependency. Koob, George, and Nora Volkow. 'Neurobiology of Addiction: A Neurocircuitry Analysis'. *The Lancet. Psychiatry* 3, no. 8 (2016): 760–73.

Overuse of digital media overstimulates attentional networks, disrupting the brain's ability to regulate focus and emotional processing, leading to heightened stress and diminished resilience. Giraldo-Luque, Santiago, Pedro Nicolás Aldana Afanador, and Cristina Fernández-Rovira. 'The Struggle for Human Attention: Between the Abuse of Social Media and Digital Wellbeing'. *Healthcare* 8, no. 4 (2020): 497; Shanmugasundaram, Mathura, and Arunkumar Tamilarasu. 'The Impact of Digital Technology, Social Media, and Artificial Intelligence on Cognitive Functions: A Review'. *Frontiers in Cognition* 2 (2023): 1203077.

Meditation, for example, strengthens our ability to observe thoughts without becoming entangled in them. This practice cultivates meta-awareness, helps us notice when deeper process thinking arises, and offers a chance to regain stability. Sandved-Smith, Lars, Casper Hesp, Jérémie Mattout, Karl Friston, Antoine Lutz, and Maxwell Ramstead. 'Towards a Computational Phenomenology of Mental Action: Modelling Meta-Awareness and Attentional Control with Deep Parametric Active Inference'. *Neuroscience of Consciousness* 2021, no. 1 (2021): niab018.

Additionally, psychoanalysis and therapeutic approaches rooted in understanding unconscious processes can help us navigate internal conflicts and provide us with a structured way to integrate the disparate layers of our mind. Mares, Leo. 'Unconscious Processes in Psychoanalysis, CBT, and Schema Therapy'. *Journal of Psychotherapy Integration* 32, no. 4 (2022): 443–52.

Seeking these resources can take many forms, from working with a therapist trained in depth psychology to exploring practices such as Internal Family Systems (IFS) or somatic therapies. 'What Is Internal Family Systems? | IFS Institute'. Accessed 23 September 2025. https://ifs-institute.com/. 'Somatic Self Care'. Accessed 23 September 2025. https://www.hopkinsmedicine.org/office-of-well-being/connection-support/somatic-self-care.

Functionally, psychedelics appear to reduce hierarchical suppression in the brain, particularly within the DMN, allowing previously constrained information to surface. This temporary loosening of rigid thought patterns enables novel perspectives and deeper emotional processing, which can lead to significant shifts in self-perception and integration of past experiences. Carhart-Harris, Robin, and Karl Friston. 'REBUS and the Anarchic Brain: Toward a Unified Model of the Brain Action of Psychedelics'. *Pharmacological Reviews* 71, no. 3 (2019): 316–44.

From an active inference perspective, psychedelics temporarily increase entropy (or free energy) in the system before helping it settle into a more stable, updated configuration. Carhart-Harris, Robin 'The Entropic Brain - Revisited'. *Neuropharmacology* 142 (2018): 167–78.

However, the effectiveness of this process depends on the set and setting and the mental framework, environment, and support structure in which the experience occurs. Elk, Michiel van, and David Bryce Yaden. 'Pharmacological, Neural, and Psychological Mechanisms Underlying Psychedelics: A Critical Review'. *Neuroscience & Biobehavioral Reviews* 140 (2022): 104793.

For example, research has shown particularly promising results for PTSD, treatment-resistant depression, and existential distress at the end of life, with studies highlighting significant reductions in symptoms for veterans and others facing deep psychological pain. Calnan, Megan, Grace Blest-Hopley, Chris Busch, et al. 'Exploring the Therapeutic Effects of Psychedelics Administered to Military Veterans in Naturalistic Retreat Settings'. *Brain and Behavior* 15, no. 7 (2025): e70660; Goodwin, Guy, Scott Aaronson, Oscar Alvarez, et al. 'Single-Dose Psilocybin for a Treatment-Resistant Episode of Major Depression'. *New England Journal of Medicine* 387, no. 18 (2022): 1637–48; Mitchell, Jennifer, Marcela Ot'alora, Bessel van der Kolk, et al. 'MDMA-Assisted Therapy for Moderate to Severe PTSD: A Randomized, Placebo-Controlled Phase 3 Trial'. *Nature Medicine* 29, no. 10 (2023): 2473–80; Rosenbaum, D., A.B. Boyle, A.M. Rosenblum, S. Ziai, M.R. Chasen, and MPhil(Pall Med). 'Psychedelics for Psychological and Existential Distress in Palliative and Cancer Care'. *Current Oncology* 26, no. 4 (2019): 225–26.

However, emerging evidence continues to suggest benefits beyond clinical diagnoses, indicating that psychedelic-assisted therapy likely also supports individuals seeking deeper personal insight and emotional resilience. Elsey, James 'Psychedelic Drug Use in Healthy Individuals: A Review of Benefits, Costs, and Implications for Drug Policy'. *Drug Science, Policy and Law* 3 (2017): 2050324517723232; Thomson, Samuel, and Nikos Thomacos. 'Utilizing Psychedelics to Enhance Well-Being: A Systematic Review'. *Journal of Psychoactive Drugs*, (2025) 1–17.

Various models exist in the literature, such as Brene Brown's BRAVING framework, which provides structure for establishing and maintaining boundaries. Brown, Brené. *Dare to Lead: Brave Work. Tough Conversations. Whole Hearts.* Random House Publishing Group, 2018.

We gain the ability to pause, reflect, and gently ask: "What's really happening here?" Gabor Maté. *The Myth of Normal: Trauma, Illness, and Healing in a Toxic Culture.* Avery/Penguin Random House, 2022.

## CHAPTER 23

These environments serve as essential foundations for neurodevelopment and shape how individuals may engage with uncertainty and belief updating. It allows us to explore, learn, and interact with the world without excessive fear or risk. Debiec, Jacek, and Regina Sullivan. 'The Neurobiology of Safety and Threat Learning in Infancy'. *Neurobiology of Learning and Memory* 143 (2017): 49–58.

This aligns with the idea that insecure or disorganized attachment styles act as 'unsafe containers,' where belief updating is constrained due to mistrust and the perceived instability of relational cues... Secure attachments, by contrast, create stable, predictable conditions that enable smooth belief updating. In a stable and supportive relationship, the child experiences predictable interactions, allowing for effective belief updating with minimal uncertainty. Cittern, David, Tobias Nolte, Karl Friston, and Abbas Edalat. 'Intrinsic and Extrinsic Motivators of Attachment under Active Inference'. *PLOS ONE* 13, no. 4 (2018): e0193955; Ainsworth, Mary, Mary Blehar, Everett Waters, and Sally Wall. *Patterns of Attachment: A Psychological Study of the Strange Situation.* Psychology Press, 2015.

This allows for more adaptive problem-solving, trust-building across teams, and greater overall organizational coherence. Balconi, Michela, Irene Venturella, Giulia Fronda, and Maria Elide Vanutelli. 'Leader-Employee Emotional "Interpersonal Tuning". An EEG Coherence Study'. *Social Neuroscience* 15, no 2 (2020): 234-243.

Internal regulation and awareness reduce tension within ourselves and make it easier for those around us to find balance, too. Stability in these moments also supports clearer communication, stronger collaboration, and more flexibility when circumstances are uncertain. Arnold, Andrew, Piotr Winkielman, and Karen Dobkins. 'Interoception and Social Connection'. *Frontiers in Psychology* 10 (2019): 2589.

On an individual level, a person who engages in self-regulation creates an internal safe container, allowing them to navigate challenges with greater resilience. Troth, Ashlea, Sandra Lawrence, Peter Jordan, and Neal Ashkanasy. 'Interpersonal Emotion Regulation in the Workplace: A Conceptual and Operational Review and Future Research Agenda'. *International Journal of Management Reviews* 20, no. 2 (2018): 523–43.

Active Inference frames these systems as niche construction, scaffolds that reduce surprise by sustaining predictability. Fabry, Regina. 'Limiting the Explanatory Scope of Extended Active Inference: The Implications of a Causal Pattern Analysis of Selective Niche Construction, Developmental Niche Construction, and Organism-Niche Coordination Dynamics'. *Biology & Philosophy* 36, no. 1 (2021): 6.

Investing too heavily in self-preservation at the expense of collaboration can lead to isolation, resulting in economic stagnation, social unrest, or even increased threats, as external entities respond with increased hostility or exclusion. Acemoglu, Daron, and James Robinson. *Why Nations Fail: The Origins of Power, Prosperity and Poverty*. Profile Books, 2013.

On the other hand, overextending without safeguards can create vulnerabilities such as erosion of core stability, dependence on unsustainable strategies, and unsustainable resource depletion. Clark, William, and Alicia Harley. 'Sustainability Science: Toward a Synthesis'. *Annual Review of Environment and Resources* 45, no. 1 (2020): 331–86; Seeliger, Leanne, and Ivan Turok. 'Towards Sustainable Cities: Extending Resilience with Insights from Vulnerability and Transition Theory'. *Sustainability* 5, no. 5 (2013): 2108–28.

As discussed in Chapter 15, somatic awareness, the ability to sense and interpret bodily signals, is central in interoception and emotional regulation. Kanbara, Kenji, and Mikihiko Fukunaga. 'Links among Emotional Awareness, Somatic Awareness and Autonomic Homeostatic Processing'. *BioPsychoSocial Medicine* 10, no. 1 (2016): 16.

From the perspective of active inference, our brains are constantly making predictions about our internal state and the external world. When we engage in somatic awareness, we refine our ability to interpret these signals accurately, reducing uncertainty and minimizing unnecessary free energy. Paulus, Martin, Justin Feinstein, and Sahib Khalsa. 'An Active Inference Approach to Interoceptive Psychopathology'. *Annual Review of Clinical Psychology* 15, no. 1 (2019): 97–122.

Self-regulation is also valuable in maintaining a safe container, as it prevents the spread of dysregulation within social interactions. Taking a deep breath before responding, grounding oneself in physical sensations, or using movement-based strategies (such as shifting posture or softening tension in the body) can help stabilize internal states. Heatherton, Todd. 'Neuroscience of Self and Self-Regulation'. *Annual Review of Psychology* 62 (2011): 363–90.

In group settings, self-regulation helps prevent emotional contagion, where dysregulated states spread and escalate uncertainty. When one person models calm, grounded presence, it provides an anchor for

Navigating Uncertainty

**others to do the same.** Troth, Ashlea, Sandra Lawrence, Peter Jordan, and Neal Ashkanasy. 'Interpersonal Emotion Regulation in the Workplace: A Conceptual and Operational Review and Future Research Agenda'. *International Journal of Management Reviews* 20, no. 2 (2018): 523–43.

**Communication is a vital yet constrained link between highly complex active inference cycles, enabling them to update and benefit from one another.** Tomasello, Michael. *Origins of Human Communication.* MIT Press, 2010.

**Effective communication, through the lens of active inference and free energy, serves as a mechanism for reducing uncertainty and fostering alignment between individuals. It involves clear, honest, and non-reactive dialogue, minimizing unnecessary cognitive and emotional friction.** Tison, Remi, and Pierre Poirier. 'Communication as Socially Extended Active Inference: An Ecological Approach to Communicative Behavior'. *Ecological Psychology* 2 (2021): 197-235; Vasil, Jared, Paul Badcock, Axel Constant, Karl Friston, and Maxwell Ramstead. 'A World Unto Itself: Human Communication as Active Inference'. *Frontiers in Psychology* 11 (2020): 417.

**In consenting contexts, supportive touch can help shift autonomic balance (often associated with lower stress responses and increased trust), which can ease belief updating and reinforce a sense of safety in relationships.** Jakubiak, Brett, and Brooke Feeney. 'A Sense of Security: Touch Promotes State Attachment Security'. *Social Psychological and Personality Science* no. 7 (2016): 745-753; Coan, James, and David Sbarra. 'Social Baseline Theory: The Social Regulation of Risk and Effort.' *Current Opinion in Psychology* 1 (2015): 87-91.

**Supportive actions include following through on commitments to build trust, respecting personal boundaries, offering help when someone is struggling while allowing them the autonomy to accept or decline, and regulating our own nervous system before engaging in emotionally charged situations. On the other hand, inconsistency, such as saying one thing and doing another, breaks trust, as does disregarding or crossing personal boundaries. Engaging in controlling behaviors, even with good intentions, can also compromise the safety of the container.** Deci, Edward, and Richard Ryan. 'The Support of Autonomy and the Control of Behavior'. *Journal of Personality and Social Psychology* 53, no. 6 (1987): 1024–37; Williams, Michele. 'Building Genuine Trust through Interpersonal Emotion Management: A Threat Regulation Model of Trust and Collaboration across Boundaries'. *The Academy of Management Review* 32, no. 2 (2007): 595–621.

**In contrast, a therapist who actively listens, responds with empathy, and allows space for emotions to be fully processed fosters a safe container. This encourages belief updating, reduces uncertainty, and supports deeper personal growth through active inference.** Holmes, Jeremy. 'Friston, Free Energy, and Psychoanalytic Psychotherapy'. *Entropy* 26, no. 4 (2024): 343.

**Attachment theory highlights how early relational experiences shape neurodevelopment, influencing how individuals process uncertainty and regulate their emotions.** Szepsenwol, Ohad, and Jeffry Simpson. 'Attachment within Life History Theory: An Evolutionary Perspective on Individual Differences in Attachment'. *Current Opinion in Psychology* 25 (2019): 65–70.

**For example, a child whose caregiver consistently responds to distress with warmth and reassurance is more likely to develop a sense of trust in others, fostering resilience in future relationships. In contrast, insecure attachments, providing inconsistent support, or responding unpredictably to a child's emotional needs, contribute to heightened uncertainty. This increases free energy in the system, reinforcing defensive behaviors that make it difficult to integrate new information.** Holmes, Jeremy. 'Friston's Free Energy Principle: New Life for Psychoanalysis?' *BJPsych Bulletin* 46, no. 3 (2022): 164–68.

**In another example, a child who experiences frequent invalidation or neglect may develop hyper-vigilance or rigid coping strategies, making belief updating and adaptation more challenging. These early experiences determine whether individuals develop epistemic trust (the confidence to accept and integrate knowledge from others) or epistemic mistrust, leading to skepticism and rigidity in thought patterns.** Fonagy, Peter, Chloe Campbell, Elizabeth Allison, and Patrick Luyten. 'Epistemic Trust and Social Learning: A Transdiagnostic Integrative Model of Mental Disorders and Therapeutic Interventions.' In Handbook of Trust and Social Psychology, edited by Kenneth Rotenberg, Serena Petrocchi, Annalisa Levante, and Flavia Lecciso. Edward Elgar, 2025. Wilkinson, Sam, Guy Dodgson, and Kevin Meares. 'Predictive Processing and the Varieties of Psychological Trauma'. *Frontiers in Psychology* 8 (2017): 1840.

**Jeremy Holmes, drawing from the Free Energy Principle, highlights that epistemic trust is essential for learning, adaptation, and psychological resilience.** Holmes, Jeremy. 'Friston's Free Energy Principle: New Life for Psychoanalysis?' *BJPsych Bulletin* 46, no. 3 (2022): 164–68.

**While we've touched on this indirectly, pausing to highlight the value of epistemic trust (the willingness to consider new knowledge as trustworthy) can be helpful, as it plays a significant yet often overlooked role in shaping how we learn and grow through relationships... Conversely, when interactions are inconsistent, dismissive, or controlling, epistemic trust is weakened, making learning and psychological growth more difficult.** Li, Elizabeth, Chloe Campbell, Nick Midgley, and Patrick Luyten. 'Epistemic Trust: A Comprehensive Review of Empirical Insights and Implications for Developmental Psychopathology'. *Research in Psychotherapy: Psychopathology, Process, and Outcome* 26, no. 3 (2023): 704.

**Fortunately, safe containers actually benefit from a lack of perfection. When small breaks in trust occur, they provide an opportunity to repair those ruptures. Addressing these moments in a timely and thoughtful way strengthens the relationship, reinforcing the understanding that it can withstand challenges like misunderstandings or inadvertent harm.** Richards, Misty, and Justin Schreiber. 'Rupture and Repair'. *Journal of the American Academy of Child and Adolescent Psychiatry* 63, no. 6 (2024): 652.

From the perspective of active inference, we do this by reducing free energy (the gap between what we expect and what we experience) through intentional actions that increase coherence and predictability. Hipólito, Inês, and Thomas van Es. 'Enactive-Dynamic Social Cognition and Active Inference'. *Frontiers in Psychology* 13 (2022): 855074.

Psychology supports this by showing how small, intentional actions can serve as regulatory anchors. This may include engaging in daily rituals, grounding through the body, naming emotions, carrying a tactile object, playing familiar calming music, or focusing on a single aspect of life that feels steady. Selvam, Raja. *The Practice of Embodying Emotions: A Guide for Improving Cognitive, Emotional, and Behavioral Outcomes*. North Atlantic Books, 2022.

Attachment theory, deeply influenced by John Bowlby and extensively articulated by Jeremy Holmes, highlights the profound significance of early relationships in shaping our lifelong emotional strategies. Holmes, Jeremy. *John Bowlby and Attachment Theory*. Routledge, 2014.

Holmes emphasizes that when caregivers sensitively respond to a child's emotional cues, they foster internal predictive models that reliably anticipate safety and support. Holmes, Jeremy. 'Friston, Free Energy, and Psychoanalytic Psychotherapy'. *Entropy* 26, no. 4 (2024): 343.

Research on animals, such as Harlow's classic experiments with monkeys, vividly illustrates the necessity of comforting, predictable presence for emotional regulation. Harlow, Harry, Margaret Kuenne Harlow, and Donald Meyer. 'Learning Motivated by a Manipulation Drive'. *Journal of Experimental Psychology* 40, no. 2 (1950): 228–34.

Securely attached individuals maintain an internal working model that predicts positive, reliable responses from significant others, effectively minimizing uncertainty and associated emotional distress. In contrast, insecure attachment, expressed as either avoidant or anxious/ambivalent, reflects maladaptive predictions about relational security. Avoidant individuals tend to expect emotional unavailability and consequently distance themselves to minimize expected disappointment. Anxious/ambivalent individuals, conversely, anticipate inconsistency, causing persistent anxiety and excessive vigilance as they attempt to achieve reassurance. Holmes, Jeremy, and Tobias Nolte. '"Surprise" and the Bayesian Brain: Implications for Psychotherapy Theory and Practice'. *Frontiers in Psychology* 10 (2019): 592.

Foster reflective functioning or mentalization—the ability to notice and understand your own emotions and the emotional states of others. Practicing mindfulness, journaling, or self-reflection helps you consciously adjust outdated or maladaptive predictive models. Bateman, Anthony, and Peter Fonagy. *Handbook of Mentalizing in Mental Health Practice*. American Psychiatric Pub, 2019.

**Holmes emphasizes the therapeutic alliance as a prime example of reshaping attachment.** Holmes, Jeremy. 'Biological v. Psychotherapeutic: Friston and Psychodynamic Therapy'. *The British Journal of Psychiatry* 209, no. 2 (2016): 171–171.


**CHAPTER 24**

**It offers a glimpse into how "othering"—the deep, often unconscious division between "us" and "them"— ultimately destabilizes everyone's active inference cycles, limiting our ability to adapt and unfold.** Tajfel, Henri, John Turner, William Austin, and Stephen Worchel. 'An Integrative Theory of Intergroup Conflict'. *Intergroup Relations: Essential Readings* (2001): 94-109.

**It's a form of "epistemic mistrust" that Peter Fonagy writes about: the immediate dismissal of information as irrelevant or untrustworthy, simply because of the source.** Fonagy, Peter, and Chloe Campbell. 'Mentalizing, Attachment and Epistemic Trust: How Psychotherapy Can Promote Resilience'. *Psychiatria Hungarica* 32, no. 3 (2017): 283–87.

**Their belief that someone from a village (or anywhere unlike their own background) is less worthy of attention is a kind of frozen narrative. [Drawing from Fonagy's work, this might be seen as "epistemic petrification": a rigid, unduly precise, internal model that can't integrate new, disconfirming data.] That rigidity prevents updating, limits exploration, and thus, limits adaptability.** Fonagy, Peter, Patrick Luyten, and Elizabeth Allison. 'Epistemic Petrification and the Restoration of Epistemic Trust: A New Conceptualization of Borderline Personality Disorder and Its Psychosocial Treatment'. *Journal of Personality Disorders* 29, no. 5 (2015): 575–609.

**As Fonagy and colleagues suggest, when people feel unseen or misrecognized, their systems learn to protect themselves by refusing incompatible inputs.** Fonagy, Peter, and Patrick Luyten. 'Attachment, Mentalizing, and the Self'. In *Handbook of Personality Disorders: Theory, Research, and Treatment, 2nd Ed.* The Guilford Press, 2018.

**Karl Friston reminds us that "we're already near Bayes-optimal," but most of our inferences run subpersonally. In other words, our brain is like an incredibly powerful and accurate machine, but it can run automatically, using outdated software.** Friston, Karl. 2024. Remarks in *Active Inference Book Project meeting*, July 11. Personal communication; Schwartenbeck, Philipp, Thomas FitzGerald, Christoph Mathys, et al. 'Optimal Inference with Suboptimal Models: Addiction and Active Bayesian Inference'. *Medical Hypotheses* 84, no. 2 (2015): 109–17.

**Specifically, Fonagy's work highlights the importance of epistemic trust. That our and others' capacity to take in new information greatly increases when it comes from someone who sees us, respects us, and aligns (at least momentarily) with our internal model.** Campbell, Chloe, and Peter Fonagy. 'Epistemic Trust and Unchanging Personal Narratives'. *Behavioral and Brain Sciences* 46 (2023).

One such fact stands out: only about 1.5% of the human genome codes for proteins — the essential molecules that form our bodies and make our cells function. The other 98.5% includes unexpressed potential. Lander, Eric, Lauren Linton, Bruce Birren, et al. 'Initial Sequencing and Analysis of the Human Genome'. *Nature* 409, no. 6822 (2001): 860–921.

Think of Wolfgang Amadeus Mozart composing symphonies as a child, Akrit Jaswal performing surgery at age seven, or Terence Tao, who scored a perfect SAT math score at eight years old. Melograni, Piero, and Lydia Cochrane. *Wolfgang Amadeus Mozart – A Biography*. University of Chicago Press, 2008; York, Carnegie Corporation of New. 'Terence Tao'. Carnegie Corporation of New York. Accessed 29 September 2025. https://www. carnegie.org/awards/honoree/terence-tao/; Zee News. 'Meet 7-Year-Old Youngest Surgeon With IQ Of 146- Know All About His Incredible Story'. Accessed 29 September 2025. https://zeenews.india.com/photos/education/ meet-7-year-old-youngest-surgeon-with-iq-of-146-know-all-about-his-incredible-story-2906534.

Neurodiversity is another fundamental characteristic of humanity. Current estimates suggest that approximately 15-20% of the population is neurodivergent in some way. This includes conditions such as autism, ADHD, dyslexia, synesthesia, and other variations in cognitive and sensory processing. Doyle, Nancy. 'Neurodiversity at Work: A Biopsychosocial Model and the Impact on Working Adults'. *British Medical Bulletin* 135, no. 1 (2020): 108–25; Singer, Judy. 'Why Can't You Be Normal for Once In Your Life? From A Problem with No Name to the Emergence of a New Category of Difference'. *Disability Discourse* (1999): 59-67.

For example, environments filled with unpredictable stimuli such as loud noises and fluctuating social cues require more processing effort, seem different, or respond in ways that may be perceived as unexpected in group settings. Chouinard, B., A. Pesquita, J. T. Enns, and C. S. Chapman. 'Processing of Visual Social-Communication Cues during a Social-Perception of Action Task in Autistic and Non-Autistic Observers'. *Neuropsychologia* 198 (2024): 108880; Cox, Cody, Lesly Krome, and Gregory Pool. 'Breaking the Sound Barrier: Quiet Spaces May Also Foster Inclusivity for the Neurodiverse Community'. *Industrial and Organizational Psychology* 17, no. 3 (2024): 350–52.

Diagnoses are best understood as models rather than rigid categories. They offer a framework for describing patterns of cognition and perception, but do not encapsulate a person's full experience or potential. Huys, Quentin, Tiago Maia, and Michael Frank. 'Computational Psychiatry as a Bridge from Neuroscience to Clinical Applications'. *Nature Neuroscience* 19, no. 3 (2016): 404–13.

From this perspective, neurodiverse cognitive processes are not 'errors' or 'deficits' but adaptive strategies—ways of perceiving and interacting with the world that expand humanity's ability to respond to different environments and internal states. Swanepoel, Annie. 'ADHD and ASD Are Normal Biological Variations as Part of Human Evolution and Are Not "Disorders"'. *Clinical Neuropsychiatry* 21, no. 6 (2024): 451–54.

**References**

This heightened awareness by the caregiver enhances nurturing environments for developing children, fostering positive outcomes like increased problem-solving approaches and increased emotional intelligence and resilience. Denham, Susanne, and Rosemary Burton. *Social and Emotional Prevention and Intervention Programming for Preschoolers*. Springer, 2003.

Neurodivergent individuals often bring heightened attention to detail, increased memory retention, and unique problem-solving approaches. Cope, Rosie, and Anna Remington. 'The Strengths and Abilities of Autistic People in the Workplace'. *Autism in Adulthood: Challenges and Management* 4, no. 1 (2022): 22–31.

For example, dyslexic thinkers frequently excel in big-picture analysis, while individuals with ADHD may thrive in dynamic, fast-paced environments, bringing fresh perspectives and adaptability to problem-solving. More broadly, a 2017 Harvard Business Review report noted that companies integrating neurodiversity can benefit from greater innovation, efficiency, and adaptability. Langston, Robert. *The Power of Dyslexic Thinking: How a Learning (Dis)Ability Shaped Six Successful Careers*. BookPros, 2009; Austin, Robert and Gary Pisano, 'Neurodiversity as a Competitive Advantage,' Harvard Business Review 95, no. 3 (2017): 96–103; Tate, Gaetana Yo. *Managing Adult ADHD in the Workplace: A Practical Guide & Workbook for Professionals*. Isohan Publishing, 2025.

Specifically, autism has been framed as a matter of precision weighting, incorporating a different balance between sensory evidence and prior expectations. [Research is still ongoing.] Individuals on the spectrum often experience heightened sensory precision, meaning their brains assign greater weight to raw sensory input rather than filtering it through prior knowledge and context. Arthur, Tom, Sam Vine, Gavin Buckingham, Mark Brosnan, Mark Wilson, and David Harris. 'Testing Predictive Coding Theories of Autism Spectrum Disorder Using Models of Active Inference'. *PLOS Computational Biology* 19, no. 9 (2023): e1011473; Pellicano, Elizabeth, and David Burr. 'When the World Becomes 'Too Real': a Bayesian Explanation of Autistic Perception.' *Trends in Cognitive Sciences* 16, no. 10 (2012): 504-510.

One way ADHD can be viewed is as a system optimized for rapid adaptation rather than sustained focus. Individuals with ADHD often demonstrate a heightened responsiveness to novelty, leading to quick shifts in attention and high energy levels. In active inference terms, their predictive models may prioritize fast-paced learning and exploration over rigid structure, making them highly adaptable in unpredictable environments. Addicott, Merideth, John Pearson, Julia Schechter, Jeffrey Sapyta, Margaret Weiss, and Scott Kollins. 'Attention-Deficit/Hyperactivity Disorder and the Explore/Exploit Trade-Off'. *Neuropsychopharmacology* 46, no. 3 (2021): 614–21.

The ability to detect patterns in chaotic environments and hyper-focus on subjects of interest can lead to groundbreaking ideas and high levels of productivity. Implementing strategies such as breaking tasks into smaller steps, using external reminders, incorporating movement into daily routines, and fostering interest-based learning enables many individuals with ADHD to enhance focus and productivity while working with their natural strengths. Lauder, Kirsty, Almuth McDowall, and Harriet Tenenbaum. 'A Systematic Review of Interventions to Support Adults with ADHD at Work—Implications from the Paucity of Context-Specific Research for Theory and Practice'. *Frontiers in Psychology* 13 (2022): 893469; Weinhardt, Justin, Ivy Mai, and Samantha Young. 'Attentional Control as a Dynamic Personal Resource: The Role of Daily ADHD Symptoms, Job Crafting, and Work Engagement'. *Journal of Business and Psychology* (2025): 1-19.

Dyslexia is often associated with difficulties in reading and language processing, but it also often comes with cognitive advantages, particularly in big-picture thinking and pattern recognition. Individuals with dyslexia tend to process information in a more holistic and associative manner rather than through linear sequencing. From an active inference perspective, their prediction models may favor global connections over fine-grained textual details, allowing them to excel in fields like engineering, design, and storytelling. Kershner, John. 'An Evolutionary Perspective of Dyslexia, Stress, and Brain Network Homeostasis'. *Frontiers in Human Neuroscience* 14 (2021): 575546.; Ziegler, Johannes, Conrad Perry, and Marco Zorzi. 'Learning to Read and Dyslexia: From Theory to Intervention Through Personalized Computational Models'. *Current Directions in Psychological Science* 29, no 3 (2020): 293–300.

While traditional literacy models can present challenges, shifting educational frameworks to embrace multimodal learning can empower dyslexic individuals to utilize their strengths, providing gifts back into society. Implementing strategies such as using audiobooks, employing speech-to-text tools, focusing on visual learning techniques, and breaking information into structured segments, many individuals with dyslexia can enhance their comprehension and learning experience while leveraging their natural strengths. Eide, Brock, and Fernette Eide. *The Dyslexic Advantage: Unlocking the Hidden Potential of the Dyslexic Brain.* Hay House, 2011.

Synesthesia is a neurological trait in which sensory inputs are automatically linked, such as seeing colors when hearing music or associating numbers with specific textures. This unique way of processing information suggests a brain that may integrate sensory signals more extensively than in neurotypical individuals. In active inference terms, some synesthetes experience a world where predictive models fuse multiple sensory modalities, potentially reducing uncertainty by reinforcing associations between disparate stimuli. Leeuwen, Tessa, Andreas Sauer, Anna-Maria Jurjut, et al. 'Perceptual Gains and Losses in Synesthesia and Schizophrenia'. *Schizophrenia Bulletin* 47, no. 3 (2021): 722–30; Seth, Anil. 'From Unconscious Inference to the Beholder's Share: Predictive Perception and Human Experience'. *European Review* 27, no. 3 (2019): 378–410.

This heightened connectivity can support enhanced memory, creativity, and artistic expression, and introduce novel ways of experiencing and interpreting information. Ward, Jamie, Daisy Thompson-Lake, Roxanne Ely, and Flora Kaminski. 'Synaesthesia, Creativity and Art: What Is the Link?' *British Journal of Psychology*

99, no. 1 (2008): 127–41; Witthoft, Nathan, and Jonathan Winawer. 'Learning, memory, and synesthesia'. *Psychological Science* 24, no. 3 (2013): 258-265.

**How would you know, and how would I? This touches on a fascinating and complex issue in philosophy of mind. [Definition: This is what cognitive scientists call the theory of mind: our capacity to infer that others have beliefs, intentions, and experiences that may differ from our own].** Brüne, Martin, and Ute Brüne-Cohrs. 'Theory of Mind—Evolution, Ontogeny, Brain Mechanisms and Psychopathology'. *Neuroscience & Biobehavioral Reviews* 30, no. 4 (2006): 437–55.

**In families and communities, neurodivergent individuals can provide unique perspectives that enhance problem-solving, deepen emotional intelligence, and create richer, more adaptive relationships.** Weinhardt, Justin, Ivy Mai, and Samantha Young. 'Attentional Control as a Dynamic Personal Resource: The Role of Daily ADHD Symptoms, Job Crafting, and Work Engagement'. *Journal of Business and Psychology* (2025): 1-19; Page, Scott E. *The Difference: How the Power of Diversity Creates Better Groups, Firms, Schools, and Societies (New Edition)*. Princeton University Press, 2008.

**This means ensuring that children today have the space and tools to develop according to their unique strengths, rather than being forced into conventional molds that may not serve them.** Grandin, Temple. *Visual Thinking: The Hidden Gifts of People Who Think in Pictures, Patterns and Abstractions*. Penguin, 2022.

### CHAPTER 26

**Given the broad spectrum of beliefs and research findings, two key principles emerge to help us come together:**

- **Religion should not be used to escalate conflict.**
- **Religious beliefs should not be used to discredit robust science. Simultaneously, those who align with science can make space for personal beliefs in others that don't fully align with scientific consensus.**

Evans, John. 'Epistemological and Moral Conflict Between Religion and Science'. *Journal for the Scientific Study of Religion* 50, no. 4 (2011): 707–27; Evans, John, and Michael Evans. 'Religion and Science: Beyond the Epistemological Conflict Narrative'. Annual Review of Sociology 34, no. 1 (2008): 87–105; Gensler, Harry. *Ethics and the Golden Rule*. Routledge, 2013.

**Both scientific and non-empirical beliefs play a crucial role in shaping how we understand and engage with the world. These perspectives (whether grounded in faith, philosophy, or empirical research) help us build shared frameworks for navigating life.** Johnson, Kathryn, Morris Okun, and Jordan Moon. 'The Interaction of Faith and Science Mindsets Predicts Perceptions of the Relationship between Religion and Science'. *Current Research in Ecological and Social Psychology* 4 (2023): 100113.

**Whether you're part of a formalized religious group like Catholicism, Judaism, Hinduism, Buddhism, or Islam, identify as a generally spiritual person, or only believe in what can be demonstrated through science, evidence points to many religious and spiritual practices offering potential benefits for humans.** Bożek, Agnieszka, Paweł F. Nowak, and Mateusz Blukacz. 'The Relationship Between Spirituality, Health-Related Behavior and Psychological Well-Being'. *Frontiers in Psychology* 11 (2020): 1997; Villani, Daniela, Angela Sorgente, Paola Iannello, and Alessandro Antonietti. 'The Role of Spirituality and Religiosity in Subjective Well-Being of Individuals With Different Religious Status'. *Frontiers in Psychology* 10 (2019): 1525; Koenig, Harold George, Dana King, and Verna Carson. *Handbook of Religion and Health*. Oup Usa, 2012.

**For others, we can look through the lens of evolutionary science, which supports the idea that humans have traditionally evolved with religious experiences as a key cornerstone of our existence.** Pyysiäinen, Ilkka, and Marc Hauser. 'The Origins of Religion: Evolved Adaptation or by-Product?' Trends in Cognitive Sciences 14, no. 3 (2010): 104–9. Boyer, Pascal. *Religion Explained: The Evolutionary Origins of Religious Thought*. Basic Books, 2007.

**An example of religion and spirituality with well-documented benefits is meditation in Buddhism, which has been shown to reduce stress and improve cognitive function.** Wang, Cheng. 'Beyond Mindfulness: How Buddhist Meditation Transforms Consciousness through Distinct Psychological Pathways'. *Frontiers in Psychology* 16 (2025): 1649564.

**While we haven't yet explored practices like prayer or chanting in Christianity and Islam, research suggests they are linked to enhanced emotional resilience and stronger social bonds.** Brandão, Tânia. 'Religion and Emotion Regulation: A Systematic Review of Quantitative Studies'. *Journal of Religion and Health* 64 (2025): 2083–100.

**From an active inference perspective, these practices provide predictable, rhythmic sensory input that helps reduce internal uncertainty and regulate emotional and interoceptive states.** Elk, Michiel van, and André Aleman. 'Brain Mechanisms in Religion and Spirituality: An Integrative Predictive Processing Framework'. *Neuroscience and Biobehavioral Reviews* 73 (2017): 359–78.

**In another similar, but distinct example, we can note that the Jewish practice of Shabbat also aligns with principles we discussed in Chapter 18, where we explored how intentional rest can restore physiological and cognitive balance, reduce internal uncertainty, and support more accurate inferences.** Cheng, Albert, Matthew Lee, and Rian Djita. 'A Cross-Sectional Analysis of the Relationship Between Sabbath Practices and US, Canadian, Indonesian, and Paraguayan Teachers' Burnout'. *Journal of Religion and Health* 62, no. 2 (2023): 1090–113; Smith-Gabai, Helene, and Ferol Ludwig. 'Observing the Jewish Sabbath: A Meaningful Restorative Ritual for Modern Times'. *Journal of Occupational Science* 18, no. 4 (2011): 347–55.

Yoga in Hinduism is yet another example—studies have shown it improves both physical and mental well-being and offers structured, embodied practices that integrate breath, movement, and focused attention. Sengupta, Pallav. 'Health Impacts of Yoga and Pranayama: A State-of-the-Art Review'. *International Journal of Preventive Medicine* 3, no. 7 (2012): 444–58.

One final example is indigenous spiritual rituals that reinforce cultural identity and emotional healing. From a scientific perspective, such practices have been linked to reduced stress, increased resilience, and stronger community ties. Elendu, Chukwuka. 'The Evolution of Ancient Healing Practices: From Shamanism to Hippocratic Medicine: A Review'. *Medicine* 103, no. 28 (2024): e39005; Koithan, Mary, and Cynthia Farrell. 'Indigenous Native American Healing Traditions'. *The Journal for Nurse Practitioners: JNP* 6, no. 6 (2010): 477–78; Xygalatas, Dimitris. 'Prediction Beyond Belief: Rituals as Active Inference Mechanisms'. *The International Journal for the Psychology of Religion* (2025): 1–5.

However, we potentially benefit most when such practices align with our understanding of active inference and the broader scientific context; otherwise, they may interfere with adaptive learning and reduce our ability to flexibly update internal models in response to new information. Aggarwal, Shilpa, Judith Wright, Amy Morgan, George Patton, and Nicola Reavley. 'Religiosity and Spirituality in the Prevention and Management of Depression and Anxiety in Young People: A Systematic Review and Meta-Analysis'. *BMC Psychiatry* 23, no. 1 (2023): 729.

In short, scientific evidence supports many religious and spiritual practices in reducing uncertainty, strengthening resilience, and supporting overall well-being□illustrating the profound role spirituality can play in enhancing human health and social cohesion. Najafi, Kazem, Hadi Khoshab, Najmeh Rahimi, and Abbas Jahanara. 'Relationship between Spiritual Health with Stress, Anxiety and Depression in Patients with Chronic Diseases'. *International Journal of Africa Nursing Sciences* 17 (2022): 100463.

Being human is incredibly complex, and finding purpose and meaning can be challenging—especially when evolutionary science describes suffering as a natural mechanism for adaptation and survival but does not assign it intrinsic meaning. Frankl, Viktor. *Man's Search For Meaning*. Random House, 2004; Reiss, Michael. 'On Suffering and Meaning: An Evolutionary Perspective'. *Modern Believing* 41, no. 2 (2000): 39–46; Pargament, Kenneth. *The Psychology of Religion and Coping: Theory, Research, Practice*. Guilford Press, 2001.

Horoscopes can serve as tools for reflection, purpose, and meaning. Noy, Shiri, Avantaea Siefke, Katie Corcoran, and Christopher Scheitle. 'Contemporary Views and Uses of Astrology in the United States: A Descriptive, Mixed-Methods Analysis'. *Social Currents* (2025).

As Carl Sagan famously said, "We are made of star-stuff." Sagan, Carl. *Carl Sagan's Cosmic Connection: An Extraterrestrial Perspective*. Edited by Jerome Agel. Cambridge University Press, 2000.

Scientists have measured fascinating and unusual cosmic phenomena like gravitational waves from colliding black holes. They've also captured neutrinos from exploding stars, mapped the invisible presence of dark matter through gravitational lensing, listened to radio pulses from spinning neutron stars (pulsars), and even imaged the shadow of a black hole using a network of telescopes spanning the globe. Cox, Brian, and Jeff Forshaw. *Black Holes: The Key to Understanding the Universe*. William Collins, 2023; Ellis, Richard S. 'Gravitational Lensing: A Unique Probe of Dark Matter and Dark Energy'. *Philosophical Transactions. Series A Mathematical, Physical, and Engineering Sciences* 368, no. 1914 (2010): 967–87; Habig, Alec T. 'Collaborative Research: SNEWS: The SuperNova Early Warning System'. NSF Award 15 (2015): 5960; Hu, Huanchen. 'Unlocking Gravity and Gravitational Waves with Radio Pulsars: Advances and Challenges'. *Astrophysics and Space Science* 370, no. 7 (2025): 74.

For example, astronomical reference points used in astrology have shifted. Earth's axis slowly wobbles over time in a cycle called axial precession, causing the zodiac constellations to drift roughly one sign every 2,000 years. *Milankovitch (Orbital) Cycles and Their Role in Earth's Climate - NASA Science*. Climate Science, 2020. https://science.nasa.gov/science-research/earth-science/milankovitch-orbital-cycles-and-their-role-in-earths-climate/.

Our cycles are naturally drawn to structures (like horoscopes) that appear to reduce uncertainty by offering explanatory narratives—tapping into our drive to minimize free energy. Bouizegarene, Nabil, Maxwell Ramstead, Axel Constant, Karl Friston, and Laurence Kirmayer. 'Narrative as Active Inference: An Integrative Account of Cognitive and Social Functions in Adaptation'. *Frontiers in Psychology* 15 (2024): 1345480.

Over time, religious interpretations shift to accommodate new understandings. Wilson, Erin K. *Religion and World Politics: Connecting Theory with Practice*. Taylor & Francis, 2022.

For example, the Ten Commandments originally implied that people were property, yet we have long since abandoned the notion that it is acceptable to own another person. Coogan, Michael. *The Ten Commandments: A Short History of an Ancient Text*. Yale University Press, 2014.

Or the Hindu practice of Sati, where widows were expected to self-immolate, erasing their agency entirely, which has been outlawed. Mani, Lata. *Contentious Traditions: The Debate on Sati in Colonial India*. University of California Press, 1998.

The caste-based restrictions in Hinduism, which also restricted role flexibility, and have been legally abolished. Goghari, Vina, and Mavis Kusi. 'An Introduction to the Basic Elements of the Caste System of India'. *Frontiers in Psychology* 14 (2023): 1210577.

**When our intentions are shaped more by societal expectations than by the genuine needs of our own systems, we often find ourselves depleted, fragmented, and disconnected. This misalignment can quietly raise free energy, leaving us out of sync with what promotes coherence and vitality.** Albarracin, Mahault, Gabriel Bouchard-Joly, Zahra Sheikhbahaee, Mark Miller, Riddhi Pitliya, and Pierre Poirier. 'Feeling Our Place in the World: An Active Inference Account of Self-Esteem'. *Neuroscience of Consciousness* 2024, no. 1 (2024): niae007.

**Similarly, heightened arousal may correspond to either excitement or anxiety, depending on contextual cues and prior expectations.** Schachter, Stanley, and Jerome Singer. 'Cognitive, Social, and Physiological Determinants of Emotional State'. *Psychological Review* 69, no. 5 (1962): 379–99; Smith, Ryan, Thomas Parr, and Karl Friston. 'Simulating Emotions: An Active Inference Model of Emotional State Inference and Emotion Concept Learning'. *Frontiers in Psychology* 10 (2019): 2844.

**Moreover, some bodily sensations can be shaped by outdated or maladaptive predictive patterns.** Castejón, Jorge, Feifan Chen, Anusha Yasoda-Mohan, Colum Ó Sé, and Sven Vanneste. 'Chronic Pain - A Maladaptive Compensation to Unbalanced Hierarchical Predictive Processing'. *NeuroImage* 297 (2024): 120711.

**Movement and exercise provide a structured opportunity to engage with and update our explanations for these sensory patterns, reducing free energy from our systems. It allows us to notice, clarify, and potentially recalibrate bodily signals.** Barca, Laura. 'The Inner Road to Happiness: A Narrative Review Exploring the Interoceptive Benefits of Exercise for Well-Being'. *Healthcare* 13, no. 16 (2025): 1960; Anderson, Michael L. 'Neural Reuse: A Fundamental Organizational Principle of the Brain', *Behavioral and Brain Sciences* 33, no. 4 (2010): 245-66.

**Just like algorithms update their models by cycling through new data, we can update our internal predictions by cycling through new experiences. Unlike algorithms, however, we do this across both mind and body.** Seth, Anil, and Manos Tsakiris. 'Being a Beast Machine: The Somatic Basis of Selfhood'. *Trends in Cognitive Sciences* 22, no. 11 (2018): 969–81.

**Take jogging: a run through a quiet park may support reflection better than one through traffic.** Bratman, Gregory, Paul Hamilton, and Gretchen Daily. 'The Impacts of Nature Experience on Human Cognitive Function and Mental Health'. *Annals of the New York Academy of Sciences* 1249 (2012): 118–36.

**Quiet mind, quiet body: Practices like meditation or breathwork train the system to detect subtle signals often drowned out in daily noise. These help reduce stress and support emotional regulation.** Fincham, Guy William, Clara Strauss, Jesus Montero-Marin, and Kate Cavanagh. 'Effect of Breathwork on Stress and Mental Health: A Meta-Analysis of Randomised-Controlled Trials'. *Scientific Reports* 13, no. 1 (2023): 432; Gibson, Jonathan. 'Mindfulness, Interoception, and the Body: A Contemporary Perspective'. *Frontiers in Psychology* 10 (2019): 2012.

**Active mind, quiet body: Activities like journaling or reading provide cognitive stimulation without sensory overload and help support symbolic reasoning and internal clarity.** Immordino-Yang, Mary Helen, and Antonio Damasio. 'We Feel, Therefore We Learn: The Relevance of Affective and Social Neuroscience to Education'. *Mind, Brain, and Education* 1, no. 1 (2007): 3–10.

**Active mind, active body: Movement that also engages strategy or coordination (like martial arts, dance, or team sports) activates real-time updating through sensorimotor integration.** Tomporowski, Phillip, and Caterina Pesce. 'Exercise, Sports, and Performance Arts Benefit Cognition via a Common Process'. *Psychological Bulletin* 145, no. 9 (2019): 929–51.

**Quiet mind, active body: Gentle activities such as walking or swimming allow the mind to soften while the body remains engaged, promoting integration and emergent insight.** Swim England, The Health and Wellbeing Benefits of Swimming: A Review (June 2017), https://www.swimming.org/swimengland/health-and-wellbeing-benefits-of-swimming/; Kelly, Paul, Chloë Williamson, Ailsa Niven, Ruth Hunter, Nanette Mutrie, and Justin Richards. 'Walking on Sunshine: Scoping Review of the Evidence for Walking and Mental Health'. *British Journal of Sports Medicine* 52, no. 12 (2018): 800–806; Biddle, Stuart JH, and Mavis Asare. 'Physical Activity and Mental Health in Children and Adolescents: a Review of Reviews.' *British Journal of Sports Medicine* 45, no. 11 (2011): 886-895.

**Research in both algorithmic training and adaptive systems shows that variation, especially when it includes unpredictable or slightly challenging conditions, can improve generalization and resilience.** Dietterich, Thomas G. 'Ensemble Methods in Machine Learning'. *In Multiple Classifier Systems.* Springer, 2000, 1–15.

**From an algorithmic perspective, this resembles training under adversarial or perturbed conditions to increase robustness and adaptability.** Goodfellow, Ian, Jonathon Shlens, and Christian Szegedy. 'Explaining and Harnessing Adversarial Examples'. arXiv:1412.6572. Preprint, arXiv, (2015); Sinha, Aman, Hongseok Namkoong, Riccardo Volpi, and John Duchi. 'Certifying Some Distributional Robustness with Principled Adversarial Training'. arXiv:1710.10571. Preprint, arXiv, (2020).

**Through the lens of active inference, our brains are constantly predicting what a sensory experience will feel like and updating those predictions based on what actually happens.** Hesp, Casper, Ryan Smith, Thomas Parr, Micah Allen, Karl Friston, and Maxwell Ramstead. 'Deeply Felt Affect: The Emergence of Valence in Deep Active Inference'. *Neural Computation* 33, no. 2 (2021): 398–446; Picard, Fabienne, and Karl Friston. 'Predictions, Perception, and a Sense of Self'. *Neurology* 83, no. 12 (2014): 1112–18.

**When we repeatedly expose ourselves to nutrient-rich or whole foods, our sensory and emotional responses begin to recalibrate. The taste becomes more rewarding as our systems learn to link it with positive, predictable outcomes, such as stable energy, improved digestion, and a grounded sense of satiation.** Khorisantono, Putu Agus, and Janina Seubert. 'How Are Food Preferences Formed and Changed? Sensory

Contributions to Anticipatory and Consummatory Processing of Food Reward'. In *Smell, Taste, Eat: The Role of the Chemical Senses in Eating Behaviour*, pp. 75-90. Edited by Lorenzo D. Stafford. Springer International Publishing, 2024.

**These updates typically take shape over days or weeks, depending on the context, history, and individual variability. But they do happen.** Friston, Karl, Thomas FitzGerald, Francesco Rigoli, Philipp Schwartenbeck, John O'Doherty, and Giovanni Pezzulo. 'Active Inference and Learning'. *Neuroscience & Biobehavioral Reviews* 68 (September 2016): 862–79; Pedraza, Felipe, Teodóra Vékony, Bence C. Farkas, et al. 'The Interplay between Executive Functions and Updating Predictive Representations'. *Scientific Reports* 15, no. 1 (2025): 30555.

## CHAPTER 28

**Our resulting perception of life—a blend of body, brain, and surroundings—is, in many ways, a hallucination.** Seth, Anil. *Being You: A New Science of Consciousness*. Penguin, 2021.

**Colors don't exist out in the world; our vibrant experience of them is entirely constructed by the brain interpreting light.** Agostini, Tiziano, Alessandra Galmonte, and Alessandro Soranzo. '13 Pictures and Color'. In *Handbook of Gestalt-Theoretical Psychology of Art*. Routledge, 2025.

**It's deeply embedded in us: our brain is wired to minimize uncertainty and exert control where possible, because doing so has historically increased our chances of survival. But in truth, we can never know with absolute certainty what's happening, only approximate the most likely explanation.** Colombo, Matteo, and Cory Wright. 'First Principles in the Life Sciences: The Free-Energy Principle, Organicism, and Mechanism'. *Synthese* 198, no. 14 (2021): 3463–88.

**The Inuit, for example, have multiple words for different types of snow, distinguishing subtle variations that an English speaker might not even notice.** Berlin, Brent, and Paul Kay. *Basic color terms: Their universality and evolution*. Univ of California Press, 1991.

**Instead, we could think of words more like subjective, derived, or individual truth. Our goal is not to find "absolute truth" but rather the best explanation that minimizes free energy (such as doubt, ambiguity, or complexity) and best fits our sensory data.** Parr, Thomas, and Giovanni Pezzulo. 'Understanding, Explanation, and Active Inference'. *Frontiers in Systems Neuroscience* 15 (2021): 772641.

**This also becomes particularly evident in politics, where uncertainty fuels strong emotions like frustration, hope, skepticism, and relief. Rather than getting swept up in these reactions, we can remember that everyone is interpreting the world through their own active inference cycles.** Kunda, Ziva. 'The Case for Motivated Reasoning.' *Psychological Bulletin* 108, no. 3 (1990): 480.

**This illustrates the idea that "all models are wrong, but some are more useful than others."** Box, George. 'Science and Statistics'. *Journal of the American Statistical Association* 71, no. 356 (1976): 791–99.

**Navigating Uncertainty**

Remember, our cycles not only absorb what's happening but have an arrow pointed in the opposite direction as well. What our cycles expect to perceive affects what we perceive. Clark, Andy. 'Whatever Next? Predictive Brains, Situated Agents, and the Future of Cognitive Science'. *Behavioral and Brain Sciences* 36, no. 3 (2013): 181–204; Hohwy, Jakob. 'Priors in Perception: Top-down Modulation, Bayesian Perceptual Learning Rate, and Prediction Error Minimization'. *Consciousness and Cognition* 47 (2017): 75–85.

This means our cognitive load increases when we encounter misleading, contradictory, or high levels of information. Kaaronen, Roope Oskari. 'A Theory of Predictive Dissonance: Predictive Processing Presents a New Take on Cognitive Dissonance'. *Frontiers in Psychology* 9 (2018): 2218; Parr, Thomas, Emma Holmes, Karl Friston, and Giovanni Pezzulo. 'Cognitive Effort and Active Inference'. *Neuropsychologia* 184 (2023): 108562.

Additionally, the effort required to process and regulate challenging emotions in these circumstances increases as well. Cancino-Montecinos, Sebastian, Fredrik Björklund, and Torun Lindholm. 'A General Model of Dissonance Reduction: Unifying Past Accounts via an Emotion Regulation Perspective'. *Frontiers in Psychology* 11 (2020): 540081.

A historical example is the witch trials that happened centuries ago, where thousands of people were accused and executed. Today, the overwhelming majority of us struggle to comprehend how this happened. However, at the time, authoritative-sounding evidence surrounded individuals, making it difficult for many to recognize how flawed the model of witches causing diseases and famine was compared to the more accurate explanations involving microbes and climate patterns. Leeson, Peter, and Jacob Russ. 'Witch Trials'. *The Economic Journal* 128, no. 613 (2018): 2066–105.


**CHAPTER 29**

I like to use the term "robust science" to describe a class of scientific inquiry that builds confidence through rigorous validation and replication across different institutions and perspectives. Munafò, Marcus R., Brian A. Nosek, Dorothy V. M. Bishop, et al. 'A Manifesto for Reproducible Science'. *Nature Human Behaviour* 1, no. 1 (2017): 0021; Nosek, Brian A., Tom E. Hardwicke, Hannah Moshontz, et al. 'Replicability, Robustness, and Reproducibility in Psychological Science'. *Annual Review of Psychology* 73, no 1 (2022): 719–48.

Consider the example of a stand-alone paper that claimed coffee consumption reduces the risk of a specific disease, only to be contradicted by later studies showing no such effect. Open Science Collaboration. ';Estimating the Reproducibility of Psychological Science.' *Science* 349, no. 6251 (2015): aac4716.

However, when multiple studies build on one another (such as through meta-analyses that aggregate findings across institutions and methodologies), we can infer higher levels of confidence. Moher, David, Alessandro Liberati, Jennifer Tetzlaff, and Douglas Altman. 'Preferred Reporting Items for Systematic Reviews and Meta-Analyses: The PRISMA Statement'. *Bmj* 339 (2009).

**References**

**Active inference has been supported by several such studies. For example, review syntheses have shown its relevance across neuroscience, artificial intelligence, and psychology.** Friston, Karl. 'Computational Psychiatry: From Synapses to Sentience'. *Molecular Psychiatry* 28, no. 1 (2023): 256–68; Langley, Christelle, Bogdan Ionut Cirstea, Fabio Cuzzolin, and Barbara J. Sahakian. 'Theory of Mind and Preference Learning at the Interface of Cognitive Science, Neuroscience, and AI: A Review'. *Frontiers in Artificial Intelligence* 5 (2022): 778852; Pezzulo, Giovanni, Thomas Parr, and Karl Friston. 'Active Inference as a Theory of Sentient Behavior'. Biological Psychology 186 (2024): 108741. Ramstead, Maxwell, Axel Constant, Paul Badcock, and Karl Friston. 'Variational ecology and the physics of sentient systems.' *Physics of life Reviews* 31 (2019): 188-205.

**From healthcare to robotics, active inference principles have been successfully applied, reinforcing its robustness.** Da Costa, Lancelot, Pablo Lanillos, Noor Sajid, Karl Friston, and Shujhat Khan. 'How Active Inference Could Help Revolutionise Robotics'. *Entropy* 24, no. 3 (2022): 361; Shusterman, Roma, Allison Waters, Shannon O'Neill, Marshall Bangs, Phan Luu, and Don Tucker. 'An Active Inference Strategy for Prompting Reliable Responses from Large Language Models in Medical Practice'. *Npj Digital Medicine* 8, no. 1 (2025): 119; Smith, Ryan, Paul Badcock, and Karl Friston. 'Recent Advances in the Application of Predictive Coding and Active Inference Models within Clinical Neuroscience'. *Psychiatry and Clinical Neurosciences* 75, no. 1 (2021): 3–13.

**Karl Friston, the pioneer of this framework and key advisor of this book, is among the most cited researchers in the world, with citations surpassing those of Einstein and Hawking.** Google Scholar. "Albert Einstein—Citations." Accessed September 25, 2025. https://scholar.google.com/citations?user=qc6CJjYAAAAJ&hl=en; Google Scholar. "Stephen Hawking—Citations." Accessed September 25, 2025. https://scholar.google.com/citations?user=-AEEg5AAAAAJ&hl=en; Research.com. "Karl Friston." Accessed September 25, 2025. https://research.com/scientists-rankings/best-scientists.

**Many modern industries are already applying principles derived from active inference, including artificial intelligence, autonomous systems, and cognitive computing.** Albarracin, Mahault, Inês Hipólito, Safae Essafi Tremblay, et al. 'Designing Explainable Artificial Intelligence with Active Inference: A Framework for Transparent Introspection and Decision-Making'. In *Active Inference*, edited by Christopher L. Buckley, Daniela Cialfi, Pablo Lanillos, et al. Springer Nature, 2024; Fox, Stephen. 'Accessing Active Inference Theory through Its Implicit and Deliberative Practice in Human Organizations'. *Entropy* 23, no. 11 (2021): 1521; Hamburg, Sarah, Alejandro Jimenez Rodriguez, Aung Htet, and Alessandro Di Nuovo. 'Active Inference for Learning and Development in Embodied Neuromorphic Agents'. *Entropy* 26, no. 7 (2024): 582; Linson, Adam, Andy Clark, Subramanian Ramamoorthy, and Karl Friston. 'The Active Inference Approach to Ecological Perception: General Information Dynamics for Natural and Artificial Embodied Cognition'. *Frontiers in Robotics and AI* 5 (2018): 21.

**Our bodies possess numerous highly effective systems to achieve this, including the traditional five senses: sight (visual system), hearing (auditory system), touch (somatosensory system), smell (olfactory**

system), and taste (gustatory system). In addition, our body sensing includes lesser-known systems balance (vestibular system), body position and movement (proprioceptive system), pain detection (nociceptive system), and temperature sensing (thermoceptive system), all working together to support our perception. Khan, Sarah, and Richard Chang. 'Anatomy of the Vestibular System: A Review'. *NeuroRehabilitation* 32, no. 3 (2013): 437–43.

The nervous system plays a crucial role, with the central nervous system (CNS) integrating sensory data with past experiences to form perceptions, and the peripheral nervous system (PNS) transmitting sensory information to the brain for interpretation. The autonomic nervous system (ANS) regulates involuntary responses, balancing the sympathetic (fight-or-flight) and parasympathetic (rest-and-digest) responses to optimize perception and readiness. Bazira, Peter. 'An Overview of the Nervous System'. *Surgery (Oxford)* 42, no. 8 (2024): 525-35.

The limbic system, including the amygdala and hippocampus, attaches emotional significance to sensory experiences, shaping how we perceive the world. Bear, Mark, Barry Connors, and Michael Paradiso. *Neuroscience: Exploring the Brain*. Jones and Bartlett Learning, 2020.

Additionally, time perception and temporal processing systems [Terms used to describe the brain's distributed mechanisms for perceiving and estimating time] help track the passage of time, allowing us to anticipate and plan actions effectively. Merchant, Hugo, Deborah Harrington, and Warren Meck. 'Neural Basis of the Perception and Estimation of Time'. *Annual Review of Neuroscience* 36 (2013): 313–36.

Emotional and social processing systems, such as the prefrontal cortex and the mirror-like neuron system, help us interpret social cues and regulate our responses, facilitating interactions and decision-making in complex social environments. Lieberman, Matthew. 'Social Cognitive Neuroscience: A Review of Core Processes'. *Annual Review of Psychology* 58 (2007): 259–89.

This process is supported by systems such as the cardiovascular system (tracking heart rate and blood pressure), the respiratory system (monitoring oxygen and carbon dioxide levels), and the endocrine system (hormonal regulation). Chen, Wen, Dana Schloesser, Angela Arensdorf, et al. 'The Emerging Science of Interoception: Sensing, Integrating, Interpreting, and Regulating Signals within the Self'. *Trends in Neurosciences* 44, no. 1 (2021): 3–16; Gordan, Richard, Judith Gwathmey, and Lai-Hua Xie. 'Autonomic and Endocrine Control of Cardiovascular Function'. *World Journal of Cardiology* 7, no. 4 (2015): 204.

The insular cortex integrates these internal signals, creating a cohesive awareness of bodily states that guide decision-making and action. Menon, Vinod. 'Insular Cortex: A Hub for Saliency, Cognitive Control, and Interoceptive Awareness'. In *Encyclopedia of the Human Brain (Second Edition)*, edited by Jordan Henry Grafman. Elsevier, 2025.

**References**

This may or may not feel like a lot of science to you [**Different manifestations of active inference prefer to make their inferences in different ways. If deep maths and science is your way, I might suggest joining the Active Inference Institute or attending the International Workshop on Active Inference**]. '6th International Workshop on Active Inference'. Accessed 25 September 2025. https://iwaiworkshop.github.io/; The Active Inference Institute. 'The Active Inference Institute'. Accessed 25 September 2025. https://www.activeinference.institute.

**CHAPTER 30**

**Large language models, autonomous vehicles, and deep learning networks usually require immense computational resources, resulting in substantial environmental footprints and power costs.** Ren, Shaolei, Bill Tomlinson, Rebecca Black, and Andrew Torrance. 'Reconciling the Contrasting Narratives on the Environmental Impact of Large Language Models'. *Scientific Reports* 14, no. 1 (2024): 26310.

**Verses takes a novel approach, applying active inference principles to enhance AI efficiency and sustainability. Verses aims to use active inference's predictive capabilities to optimize how AI processes information, significantly reducing computational demands.** VERSES. 'Imagine a Smarter World'. Accessed 28 September 2025. https://www.verses.ai.

**At Cortical Labs, the blend between biological neurons and digital computing is becoming a reality. Their "DishBrain" project is an audacious yet elegant demonstration of active inference in action, using real neurons to learn, adapt, and perform computational tasks.** Cortical Labs. 'Cortical Labs'. Accessed 28 September 2025. https://corticallabs.com. DeMarse, Thomas B., Daniel A. Wagenaar, Axel W. Blau, and Steve M. Potter. 'The Neurally Controlled Animat: Biological Brains Acting with Simulated Bodies.' *Autonomous Robots* 11, no. 3 (2001): 305-310.

**DishBrain employs a fusion of biological neurons and silicon interfaces, harnessing active inference principles to enable living neurons to predict and respond to digital stimuli.** Kagan, Brett, Forough Habibollahi, Brad Watmuff, et al. 'Harnessing Intelligence from Brain Cells In Vitro'. *The Neuroscientist* 31, no. 5 (2025): 536–55.

**Such bio-computational devices offer the potential for remarkable adaptability, learning efficiency, and resilience compared to traditional silicon-only technologies. For instance, biomimetic-computational devices like DishBrain, neuromorphic computing, photonics, and memristors could revolutionize personalized medicine by dynamically adapting to an individual's unique biological responses, optimizing treatments in real-time.** Kagan, Brett, Christopher Gyngell, Tamra Lysaght, Victor Cole, Tsutomu Sawai, and Julian Savulescu. 'The Technology, Opportunities, and Challenges of Synthetic Biological Intelligence'. *Biotechnology Advances* 68 (2023): 108233; Watmuff, Bradley, Forough Habibollahi, Candice Desouza, et al. 'Drug Treatment Alters Performance in a Neural Microphysiological System of Information Processing'. *Communications Biology* 8, no. 1 (2025): 916.

However, computational methods grounded in active inference could systematically incorporate affect-like signals [akin to emotional and bodily sensations], analyzing them objectively rather than reacting impulsively as humans often do under stress. When analytical tools are used alongside human emotional and bodily insight, governance has the potential to become significantly strengthened, especially in emotionally charged decision making scenarios. Hesp, Casper, Ryan Smith, Thomas Parr, Micah Allen, Karl Friston, and Maxwell Ramstead. 'Deeply Felt Affect: The Emergence of Valence in Deep Active Inference'. *Neural Computation* 33, no. 2 (2021): 398–446; Tsakiris, Manos, Neza Vehar, and Raffaele Tucciarelli. 'Visceral Politics: A Theoretical and Empirical Proof of Concept'. Philosophical Transactions of the Royal Society of London. *Series B, Biological Sciences* 376, no. 1822 (2021): 20200142.

Consider the critical decisions following the September 11th attacks in the United States. These decisions led to prolonged military engagements where anticipated weapons of mass destruction were ultimately not found, contributing to substantial financial debt and tragic loss of life. Dearstyne, Bruce. 'The FDNY on 9/11: Information and Decision Making in Crisis'. *Government Information Quarterly* 24, no. 1 (2007): 29–46.

Crucially, however, AI alone cannot dictate ethical choices—it needs to be guided by human-defined values. These values could be determined in calm, reflective states, ensuring that computational tools provide meaningful clarity precisely when we most need it. De Cremer, David, and Devesh Narayanan. 'How AI Tools Can—and Cannot—Help Organizations Become More Ethical'. *Frontiers in Artificial Intelligence* 6 (2023): 1093712; Floridi, Luciano, and Josh Cowls. 'A Unified Framework of Five principles for AI in Society.' *Machine Learning and the City: Applications in Architecture and Urban Design* (2022): 535–545.

By modeling an individual's cognitive patterns and psychological states through active inference frameworks, computational psychiatry can provide highly personalized and targeted therapeutic interventions. Friston, Karl. 'Computational Psychiatry: From Synapses to Sentience'. *Molecular Psychiatry* 28, no. 1 (2023): 256–68; Friston, Karl. 'Precision Psychiatry'. *Biological Psychiatry. Cognitive Neuroscience and Neuroimaging* 2, no. 8 (2017): 640–43.

This targeted approach signals a breakthrough in mental health treatments, from anxiety and depression to more complex psychological conditions. When therapists work through the lens of active inference, they're better able to align treatment paths that reflect each patient's individual experience. Krupnik, Valery. 'The Therapeutic Alliance as Active Inference: The Role of Trust and Self-Efficacy'. *Journal of Contemporary Psychotherapy* 53, no. 3 (2023): 207–15.

Together, these examples highlight the transformative opportunity [While not explored in depth here, governance is another domain where active inference could drive meaningful change—enabling institutions to function more like living systems: sensing, predicting, and dynamically responding to the evolving needs of their people. By embracing uncertainty and continuously updating their internal

**References**

models, governments could become more adaptive, resilient, and attuned to the needs of the people.] of active inference beyond the focused application of this book. Deslatte, Aaron, Jeffery A. Adams, Faisal Cheema, Jesse L. Barnes, Elizabeth A. Koebele, and Sara Alonso Vicario. 'Understanding the Impact of Institutions on Climate-Adaptive Policy Designs: A Study of Collective Action Inference in Urban Water Systems'. *Policy Studies Journal* 53, no. 3 (2025): 637–53.

**CHAPTER 31**

**Breathwork: consider breathwork that might serve your system or help soften underlying tension.** Fincham, Guy William, Clara Strauss, Jesus Montero-Marin, and Kate Cavanagh. 'Effect of Breathwork on Stress and Mental Health: A Meta-Analysis of Randomised-Controlled Trials'. *Scientific Reports* 13, no. 1 (2023): 432.

**Nature Immersion: consider getting out into nature in a way that supports your system or reveals subtle dissonance worth exploring.** Cox, Daniel, Danielle Shanahan, Hannah Hudson, et al. 'Doses of Nearby Nature Simultaneously Associated with Multiple Health Benefits'. *International Journal of Environmental Research and Public Health* 14, no. 2 (2017): 172.

**Concept Mapping: A visual method for organizing ideas and making relationships between them visible.** Trochim, William, and Daniel McLinden. 'Introduction to a Special Issue on Concept Mapping'. *Evaluation and Program Planning* 60 (2017): 166–75.

**Stretching or Yoga: Try choosing a few slow, mindful movements (like a gentle twist, forward fold, or shoulder roll) and notice how your body feels before, during, and after.** Zvetkova, Elissaveta, Eugeni Koytchev, Ivan Ivanov, Sergey Ranchev, and Antonio Antonov. 'Biomechanical, Healing and Therapeutic Effects of Stretching: A Comprehensive Review'. *Applied Sciences* 13, no. 15 (2023): 8596.

**Curating Workspace for Sensory Ease: Take a few minutes to look around your workspace and notice what feels calming or overstimulating.** Wells, Meredith. 'Office Clutter or Meaningful Personal Displays: The Role of Office Personalization in Employee and Organizational Well-Being'. *Journal of Environmental Psychology* 20, no. 3 (2000): 239–55.

**Free energy arises when expectations clash with reality. These mismatches serve as signals rather than faults. They offer opportunities to pause, reflect, and adjust. Because ultimately, your system is built for this. Your body and brain are built to respond to mismatches and are continuously recalibrating.** Friston, Karl. 'The Free-Energy Principle: A Unified Brain Theory?' *Nature Reviews Neuroscience* 11, no. 2 (2010): 127–38.

**Thought labeling: Briefly notice and name the type of thought you're having (e.g., worry, planning, judging) to create a bit of space between you and the thought.** Deacon, Brett, Tamer Fawzy, James Lickel, and Kate Wolitzky-Taylor. 'Cognitive Defusion versus Cognitive Restructuring in the Treatment of Negative Self-Referential

Thoughts: An Investigation of Process and Outcome'. *Journal of Cognitive Psychotherapy* 25, no. 3 (2011): 218–32.

**Progressive muscle relaxation: Gently tense and then release one muscle group at a time (from your toes to your face) to support body awareness and relaxation.** Muhammad Khir, Syazwina, Wan Mohd Azam Wan Mohd Yunus, Norashikin Mahmud, et al. 'Efficacy of Progressive Muscle Relaxation in Adults for Stress, Anxiety, and Depression: A Systematic Review'. *Psychology Research and Behavior Management* 17 (2024): 345–65.

**Sensory deprivation tank: Spend time in a quiet, dark float tank to reduce sensory input and notice how your attention and awareness shift in the absence of external stimuli.** Lashgari, Elnaz, Emma Chen, Jackson Gregory, and Uri Maoz. 'A Systematic Review of Flotation-Restricted Environmental Stimulation Therapy (REST)'. *BMC Complementary Medicine and Therapies* 25 (2025): 230.

**Body scan meditation: Slowly bring attention to different parts of your body (from head to toe), pausing at each one to notice sensation without judgment.** Gan, Ruochen, Liuyi Zhang, and Shulin Chen. 'The Effects of Body Scan Meditation: A Systematic Review and Meta-Analysis'. *Applied Psychology: Health and Well-Being* 14, no. 3 (2022): 1062–80.

**Socratic questioning: Ask yourself open-ended questions like "What else might be true?" or "Is this just a familiar thought pattern?" to explore and gently challenge assumptions.** Clark, Gavin, and Sarah Egan. 'The Socratic Method in Cognitive Behavioural Therapy: A Narrative Review'. *Cognitive Therapy and Research* 39, no. 6 (2015): 863–79.

**Joyful exploration, far from a frivolous luxury, offers a highly effective way to reduce uncertainty and learn what's possible. It helps systems recover from stress, expand tolerance windows, and develop creative responses to novel situations. Group play, in particular, supports co-regulation, builds trust, and lowers shared uncertainty, benefiting not just individuals but whole groups.** Porges, Stephen, and Deb Dana. *Clinical Applications of the Polyvagal Theory: The Emergence of Polyvagal-Informed Therapies.* W. W. Norton & Company, 2018.

**Willpower relies on top-down override; volition arises from system-wide integration. When your body, environment, and values are in sync, you don't need to push in the same way, and effort softens.** Tang, Yi-Yuan, Rongxiang Tang, Michael Posner, and James Gross. 'Effortless Training of Attention and Self-Control: Mechanisms and Applications'. *Trends in Cognitive Sciences* 26, no. 7 (2022): 567–77.

**It allows your system to sort signal from noise, discern what truly matters, and recalibrate without the overwhelm of constant input. You don't stop learning when you pause.** Jebelli, Joseph. The Brain at Rest: Why Doing Nothing Can Change Your Life. Torva, 2025; Raichle, Marcus. 'The Brain's Default Mode Network.' *Annual Review of Neuroscience* 38, no. 1 (2015): 433-447.

**Understanding your system reduces unnecessary free energy. When you can anticipate your own tendencies, triggers, and needs, your system doesn't waste energy managing surprise. Instead, it can reallocate that energy toward growth, recalibration, or intentional response.** Bauermeister, Jonathan, and Pablo Lanillos. 'The Role of Valence and Meta-Awareness in Mirror Self-Recognition Using Hierarchical Active Inference'. In *Active Inference*, pp. 112-129 Springer Nature, 2023.

**Walking meditation: Walk slowly and with intention, paying attention to the sensations in your feet, breath, and surroundings.** Teut, M., E. J. Roesner, M. Ortiz, et al. 'Mindful Walking in Psychologically Distressed Individuals: A Randomized Controlled Trial'. *Evidence-Based Complementary and Alternative Medicine* 2013, no. 1 (2013): 489856.

**Through the lens of active inference, setting intention helps reduce prediction error, promoting clarity and coherence across your system.** Safron, Adam. 'The Radically Embodied Conscious Cybernetic Bayesian Brain: From Free Energy to Free Will and Back Again'. *Entropy* 23, no. 6 (2021): 783.

**One possible response is naming the lack of safety without judgment. Simply saying, "I don't feel safe right now," can begin to reengage the prefrontal cortex and soften reactive patterns. Another response is to seek partial safety. Even small anchors, such as a steady breath, a grounding texture, a calm environment, or the steady presence of someone you trust, can signal enough safety for the system to begin settling.** Porges, Stephen. 'Polyvagal Theory: A Science of Safety'. *Frontiers in Integrative Neuroscience* 16 (2022): 871227; Punkanen, Marko, and Tony Buckley. 'Embodied Safety and Bodily Stabilization in the Treatment of Complex Trauma'. *European Journal of Trauma & Dissociation* 5, no. 3 (2021): 100156.

**Self-compassion meditation: Sit quietly and bring to mind a gentle phrase (like "May I hold myself with compassion") while noticing any sensations or emotions that arise.** Augusta Quist Møller, Selma A., Sohrab Sami, and Shauna Shapiro. 'Health Benefits of (Mindful) Self-Compassion Meditation and the Potential Complementarity to Mindfulness-Based Interventions: A Review of Randomized-Controlled Trials'. *OBM Integrative and Complementary Medicine* 4, no. 1 (2019): 1–20; Neff, Kristin. 'Self-Compassion: An Alternative Conceptualization of a Healthy Attitude Toward Oneself.' *Self and Identity* 2, no. 2 (2003): 85-101.

**Partnered breath synchronization: Sit with a trusted partner and gently match the rhythm of your breath with theirs, perhaps while sitting intertwined, to support connection and calm.** Balconi, Michela, and Laura Angioletti. 'Social Interoception and Autonomic System Reactivity during Synchronization Behavior'. *Behavioral Sciences* 14, no. 3 (2024): 149; Coutinho, Joana F., Markku Penttonen, Anu Tourunen, et al. 'Electrodermal and Respiratory Synchrony in Couple Therapy in Distinct Therapeutic Subsystems and Reflection Periods'. *Psychotherapy Research: Journal of the Society for Psychotherapy Research* 35, no. 2 (2025): 207–22.

**Free energy isn't always a problem. Sometimes, it's what sparks transformation and fuels curiosity,**

**Navigating Uncertainty**

**movement, and entirely new ways of seeing.** Friston, Karl, Marco Lin, Christopher Frith, Giovanni Pezzulo, Allan Hobson, and Sasha Ondobaka. 'Active Inference, Curiosity and Insight'. *Neural Computation* 29, no. 10 (2017): 2633–83.

**Cold plunge: Briefly immerse your body in cold water or an ice bath to practice sitting with an activated nervous system—notice your breath, sensations, and how your state shifts afterward.** Reed, Emma L., Christopher L. Chapman, Emma K. Whittman, et al. 'Cardiovascular and Mood Responses to an Acute Bout of Cold Water Immersion'. *Journal of Thermal Biology* 118 (2023): 103727.

**Connection enhances prediction accuracy. Being in close, trusted relationships helps regulate each person's nervous system and sharpens the ability to interpret social and environmental cues accurately.** Molapour, Tanaz, Cindy Hagan, Brian Silston, et al. 'Seven Computations of the Social Brain'. *Social Cognitive and Affective Neuroscience* 16, no. 8 (2021): 745–60.

**Loving-kindness meditation: Silently repeat simple phrases like "May I be safe, may I be well," first toward yourself, then toward others.** Zeng, Xianglong, Cleo Chiu, Rong Wang, Tian Oei, and Freedom Leung. 'The Effect of Loving-Kindness Meditation on Positive Emotions: A Meta-Analytic Review'. *Frontiers in Psychology* 6 (2015): 1693.

**Groups that embrace a variety of internal models tend to explore more possibilities, tolerate more uncertainty, and generate more innovative solutions.** Page, Scott. *The Difference: How the Power of Diversity Creates Better Groups, Firms, Schools, and Societies.* Princeton University Press, 2008.

**Contact improvisation: Join a class or explore gentle movement with a partner, staying present and responsive to each other's touch, pressure, and direction.** Purr, Ariane, Wim Waterink, and Susan van Hooren. 'The Effect of Touch on Affect, Stress, Sense of Connectedness and Sense of Self: An Experimental Study on Contact Improvisation Dance'. *The Arts in Psychotherapy* 93 (2025): 102272.

**Some personal favorites of mine include slow jogging through my neighborhood park while listening to a Yielding playlist by my friend Jeff Krone, unwinding in the evening with a book by Elizabeth Strout or a Kurzgesagt video, and exploring insights on active inference from Karl Friston, Andy Clark, Anil Seth, and Lisa Feldman Barrett while working on a puzzle.** Barrett, Lisa Feldman. *How Emotions Are Made: The Secret Life of the Brain.* Macmillan, 2017; Clark, Andy. *Surfing Uncertainty: Prediction, Action, and the Embodied Mind.* Oxford University Press, 2016; Parr, Thomas, Giovanni Pezzulo, and Karl Friston. Active Inference: The Free Energy Principle in Mind, Brain, and Behavior. The MIT Press, 2022; Seth, Anil. *Being You: A New Science of Consciousness.* Faber & Faber, 2022.

CHAPTER 32

**The act of birth itself can be experienced as traumatic, a profound adjustment to life's uncertainties.** Hanh, Thich Nhat. *Fear: Essential Wisdom for Getting Through the Storm.* Random House, 2012; Sun, Xiaoqing, Xuemei Fan, Shengnan Cong, et al. 'Psychological Birth Trauma: A Concept Analysis'. *Frontiers in Psychology* 13 (2023): 1065612.

**Choosing intentional ways to shift the ripples in ourselves from previous generations (who were doing their best under their circumstances and resources) can support breaking patterns that might otherwise repeat across generations, allowing for more ease, adaptability, and resilience across time.** Yehuda, Rachel, Nikolaos Daskalakis, Linda Bierer, Heather Bader, Torsten Klengel, Florian Holsboer, and Elisabeth Binder. "Holocaust Exposure Induced Intergenerational Effects on FKBP5 Methylation.' *Biological Psychiatry* 80, no. 5 (2016): 372-380.

**In the end, overcoming hardship can build an incredibly valuable life skill, useful not just for personal well-being, but also as a beneficial asset in work culture, innovation, and genuine connection.** Henson, Charlotte, Didier Truchot, and Amy Canevello. 'What Promotes Post Traumatic Growth? A Systematic Review'. *European Journal of Trauma & Dissociation* 5, no. 4 (2021): 100195; Bonanno, George. 'Loss, Trauma, and Human Resilience: Have We Underestimated the Human Capacity to Thrive After Extremely Aversive Events?.' *American Psychologist* 59, no. 1 (2004): 20.

CHAPTER 33

**What looks like inaction is sometimes a deliberate response that offers preservation, integration, or a chance to recalibrate. Just as rest can refine our internal models, and spaciousness gives rise to clarity, choosing not to act can be an intentional and powerful form of action.** Finn, Emily. 'Is It Time to Put Rest to Rest?' *Trends in Cognitive Sciences* 25, no. 12 (2021): 1021–32; Raichle, Marcus E. 'The Brain's Default Mode Network.' *Annual Review of Neuroscience* 38, no. 1 (2015): 433-447.

**Drawing from Howard Gardner's Theory of Multiple Intelligences, we can appreciate that different people have different domains in which they thrive: linguistic, musical, spatial, kinesthetic, interpersonal, intrapersonal, naturalistic, and beyond.** Gardner, Howard. *Intelligence Reframed: Multiple Intelligences for the 21st Century.* Basic Books, 1999.

**Intentional communities act as both a buffer and a catalyst. They reduce uncertainty through shared meaning, reinforce adaptive models through feedback, and offer emotional scaffolding when we face challenges.** Albarracin, Mahault, Daphne Demekas, Maxwell Ramstead, and Conor Heins. 'Epistemic Communities under Active Inference'. *Entropy* 24, no. 4 (2022): 476; Michalski, Camilla, Lori Diemert, John Helliwell, Vivek Goel, and Laura Rosella. 'Relationship between Sense of Community Belonging and Self-Rated Health across Life Stages'. *SSM - Population Health* 12 (2020): 100676.

**No matter what your current call to action is, no matter the circumstances you were born into, the truth remains: suffering is not relative. We all suffer.** Aich, Tapas Kumar. 'Buddha Philosophy and Western Psychology'. *Indian Journal of Psychiatry* 55, no. Suppl 2 (2013): S165–70; Frankl, Viktor. *Man's Search For Meaning*. Random House, 2004.

# INDEX

**A**

**Accuracy:** How well our internal models fit our sensory perceptions.

**Active inference cycle:** A scientific concept used to describe how our brains continuously sense and interpret the world around us, test those interpretations through our actions, and then adjust based on the feedback we receive from our environment, body, and mind.

**Allostasis:** Adaptive ability that enables us to move through difficulty and return to center.

**Attentional narrowing:** A natural brain response driven by stress and uncertainty.

**B**

**Bidirectional framework:** Functioning in two directions.

**Biomarkers:** Objective signals—like heart rate or hormone levels—that reflect what's happening in the body at a specific moment in time.

**C**

**Computational diversity** A metaphor for the different ways humans minimize uncertainty, process sensory input, and engage with the world.

**Complexity:** The degree of detail in our models.

**D**

**Dehumanization:** The process of depriving a person or group of positive human qualities.

**E**

**Embodiment:** The body takes an active role in shaping how we think, feel, perceive, and act, rather than merely housing the mind.

**Epistemic forging:** Active exploration driven by curiosity to reduce uncertainty and gain understanding

**Epistemic petrification:** a rigid, unduly precise, internal model that struggles to integrate new, disconfirming data.

**Epistemic trust:** The willingness to consider new knowledge as trustworthy.

**Exploration:** Involves seeking out new experiences or information to expand your understanding of the world.

**F**

**Free energy principle:** Every action, thought, and feeling is part of a larger effort to reduce surprise and uncertainty in your life. Its function is to align your internal expectations with the world around you.

**H**

**Homeostasis:** A quiet, behind-the-scenes process that keeps us alive and steady.

**I**

**Insecure attachment:** Expressed as either avoidant or anxious/ambivalent, reflects maladaptive predictions about relational security.

**Interoceptive inference:** Using bodily sensations with cognitive signals to maintain balance and a coherent sense of self.

**M**

**Markov blanket:** The boundary that enables a system to interact with its environment while preserving its own integrity.

**N**

**Neural gain:** The amplification of strong signals and suppression of weaker ones.

**P**

**Policies:** Possible actions.

**Precision:** The confidence we place in a specific signal or piece of information.

**Priors:** Pre-existing beliefs.

**Primary process:** Associated with instinctual, emotion-driven thought (the id).

**Pragmatic:** What is already known.

**R**

**Regulatory Blindness:** A state where someone appears steady in one area but misses how their actions create instability in another. In Active Inference terms, it reflects misjudging which signals to trust, so regulation works locally but fails more broadly.

**S**

**Safe container:** Any environment (physical, emotional, or relational) that provides stability, predictability, and a sense of security within our active inference cycles.

**Secure attachment:** Individuals maintain an internal working model that predicts positive, reliable responses from significant others, effectively minimizing uncertainty and associated emotional distress.

**Secondary process:** Secondary, logical, structured, and tied to our sense of self (the ego).

**Self-efficacy:** The belief in one's ability to execute actions and influence outcomes.

**Self-evidencing:** Process by which our brains continuously seek to maximize the evidence for their internal models of the world. Essentially, the brain is constantly testing its predictions against sensory input to ensure that its understanding of the environment is as accurate and useful as possible.

**Self-knowing:** An ongoing process of making sense of how your inner world and the outer world interact.

**Sense-making process:** The process by which people give meaning to their collective experiences.

**Sensory attenuation:** The brain filtering out sensory inputs during action, temporarily ignoring the immediate sensory feedback of our actions.

**Somatic awareness:** The ability to sense and interpret bodily signals, central in interoception and emotional regulation.

**Synaptic gain:** This process dynamically balances neural excitation and inhibition, prioritizing the most relevant information for action and perception.

**Synaptic homeostasis:** A process involving pruning unnecessary neural connections while also strengthening those most relevant to learning and memory, leaving behind a refined, simplified structure that is better suited for the challenges of the next day.

**Systems thinking:** A way of making sense of the world's complexity by viewing it in terms of wholes, patterns, and relationships—rather than breaking things down into isolated parts.

## T

**Theory of mind:** Our capacity to infer that others have beliefs, intentions, and experiences that may differ from our own.

**Truth is provisional:** Existing as our best understanding at this moment—open to refinement and evolution as new insights or experiences emerge.

## U

**Utilization (exploitation):** Involves using existing knowledge to navigate familiar situations effectively.

## V

**Volition:** The ability to align our actions with our beliefs and goals.